KNOW SOIL KNOW LIFE

KNOW SOIL KNOW LIFE

DAVID L. LINDBO, DEB A. KOZLOWSKI, AND CLAY ROBINSON, *Editors*

COPYRIGHT © 2012 SOIL SCIENCE SOCIETY OF AMERICA

ALL RIGHTS RESERVED. No part of this publication may be reproduced or transmitted in any form or by any means, electronic or mechanical, including photocopying, recording, or any information storage and retrieval system, without permission in writing from the publisher.

The views expressed in this publication represent those of the individual Editors and Authors. These views do not necessarily reflect endorsement by the Publisher(s). In addition, trade names are sometimes mentioned in this publication. No endorsement of these products by the Publisher(s) is intended, nor is any criticism implied of similar products not mentioned.

Soil Science Society of America
5585 Guilford Road, Madison, WI 53711-5801 USA
608-273-8080 | soils.org

dl.sciencesocieties.org
SocietyStore.org

ISBN: 978-0-89118-954-1 (print)
ISBN: 978-0-89118-955-8 (electronic)
doi:10.2136/2012.knowsoil

Library of Congress Control Number: 2012954713
ACSESS Publications
ISSN 2165-9834 (print)
ISSN 2165-9842 (online)

ASA, CSSA, and SSSA Book and Multimedia Publishing Committee
  April Ulery, Chair
  Warren Dick, ASA Editor-in-Chief
  E. Charles Brummer, CSSA Editor-in-Chief
  Andrew Sharpley, SSSA Editor-in-Chief
  Mary Savin, ASA Representative
  Mike Casler, CSSA Representative
  David Clay, SSSA Representative
  Lisa Al-Amoodi, Managing Editor

Cover and interior design: Kristyn Kalnes
Cover image: USDA-NRCS
Christine Mlot, developmental editor
John Lambert, custom illustrations

Printed in the United States of America.

KNOW SOIL KNOW LIFE

To all our introductory soil science professors.
You helped us see soil
as something other than dirt.

To all our students, past, present, and future,
who we hope will come to view soil with awe and fascination.

"Heaven is under our feet as well as over our heads."
Henry David Thoreau

CONTENTS

Preface, ix
Acknowledgments, x
Author Biographies, xi

CHAPTER 1
KNOW SOIL, KNOW LIFE, 1
by David Lindbo, John Havlin, Deb Kozlowski, and Clay Robinson

SOIL'S IMPORTANCE TO PEOPLE, **1**
SOIL'S IMPORTANCE IN THE ENVIRONMENT, **2**
HOW MUCH SOIL IS ON EARTH, **4**
WHAT IS SOIL, **6**
BASIC SOIL PROCESSES, **6**
SOILS AND BIOMES, **9**
POPULATION GROWTH AND SOIL, **9**
SUMMARY, **13**

CHAPTER 2
PHYSICAL PROPERTIES OF SOIL AND SOIL FORMATION, 15
by David Lindbo, Wale Adewunmi, and Rich Hayes

SOIL COLOR, **15**
SOIL TEXTURE, **17**
SOIL STRUCTURE, **24**
SOIL CONSISTENCE, **26**
SOIL HORIZONS AND PROFILE, **27**
BULK DENSITY, **29**
WATER AND WATER MOVEMENT, **29**
DEPTH TO SEASONAL SATURATION, **31**
PERMEABILITY, **32**
INFILTRATION, **32**
SOIL ORGANIC MATTER, **32**
SOIL FORMATION, **33**
SUMMARY, **45**

CHAPTER 3
SOIL BIOLOGY: THE LIVING COMPONENT OF SOIL, 49
by Tom Loynachan

ORGANISMS OF THE SOIL, **51**
LIFE ABOVE THE SOIL DEPENDS ON LIFE IN THE SOIL, **57**
SOIL QUALITY AND SOIL MANAGEMENT, **61**
SUMMARY, **66**

CHAPTER 4
CHEMICAL PROPERTIES OF SOIL: SOIL FERTILITY AND NUTRIENT MANAGEMENT, 69
by John Havlin and Bianca Moebius-Clune

NUTRIENT SUPPLY TO PLANTS AND NUTRIENT CYCLING IN SOIL, **70**
NUTRIENT MOBILITY IN SOIL, **72**
SOIL PROPERTIES INFLUENCING NUTRIENT SUPPLY TO PLANTS, **73**
ASSESSING NUTRIENT DEFICIENCY IN PLANTS, **76**
NUTRIENT MANAGEMENT, **79**
SUMMARY, **80**

CHAPTER 5
SOIL CLASSIFICATION, SOIL SURVEY, AND INTERPRETATIONS OF SOIL, 83
by David Lindbo, Doug Malo, and Clay Robinson

SOIL CLASSIFICATION, **83**
SOIL SURVEYS AND MAPPING, **92**
INTERPRETATIONS, **97**
SUMMARY, **105**

CHAPTER 6
ENVIRONMENTAL SCIENCE, SOIL CONSERVATION, AND LAND USE MANAGEMENT, 109
by Clay Robinson, Wale Adewunmi, David Lindbo, and Bianca Moebius-Clune

NATURAL PROCESSES AFFECTING SOIL DEGRADATION, **113**
HUMAN ACTIVITIES AFFECTING SOIL DEGRADATION, **128**
SUMMARY, **136**

CHAPTER 7
SOILS AND BIOMES, 139
by Clay Robinson, Wale Adewunmi, and David Lindbo

FORESTS, **139**
GRASSLANDS, **148**
TUNDRA, **152**
DESERTS, **154**
SHRUBLANDS, **157**
AQUATIC BIOMES/WETLANDS, **158**
SUMMARY, **161**

CHAPTER 8
SOILS AND SOCIETY, 163
by Melanie Szulczewski, Mandy Liesch, John Havlin, and David Lindbo

SOILS AND HUMAN CULTURE, **163**
SOILS AND HUMAN HEALTH, **168**
CHALLENGES TO THE SOIL: THEN, NOW, AND LOOKING AHEAD, **169**
HAVE WE LEARNED ENOUGH FROM PAST MISTAKES? **179**
SUMMARY, **181**

CHAPTER 9
CAREERS IN SOIL SCIENCE: DIG IN, MAKE A DIFFERENCE, 183
by Susan Chapman and David Lindbo

SO THEN, WHAT DOES A SOIL SCIENTIST DO? **184**
WHO HIRES A SOIL SCIENTIST? **184**
WHAT TYPE OF EDUCATION DO YOU NEED? **184**
DO I NEED A CERTIFICATION OR A LICENSE
    TO BE A SOIL SCIENTIST?, **185**
A SERIES OF ADVENTURES!
    LISA "DIRTGIRL" BRYANT, **186**
A LIFELONG SCIENTIST!
    BILL SHUSTER, **188**
MY DIVERSE LIFE AS A SOIL SCIENTIST.
    DAWN FERRIS, **190**
NEVER STOP LEARNING!
    MISSY HOLZER, **192**
ENJOY THE CHALLENGE!
    MATTHEW "MATT" DUNCAN, **194**
SOLVING PROBLEMS, *RECLAIMING LAND*.
    BRUCE BUCHANAN, **196**

CHAPTER 10
SUMMARY AND PERSPECTIVES, 199
by David Lindbo and Deb Kozlowski

PREFACE

You are about to begin a tour of an often overlooked part of our natural world—the soil. True, it does not have the immediate appeal of the cute, cuddly creatures that live in the forests or the power of a volcano or the vastness of space, yet soils have an inner beauty of their own. This book will expose you to the nature of soils and soil science. You will see that the stuff you probably call dirt (and please do not use that word again!) is just as complicated, interesting, and varied as any other part of our world. We have distilled a complex science down to about 200 pages. This writing is geared towards a young adult audience. High school students studying environmental science or participating in Envirothon or Science Olympiad will have an easily accessible resource. Undergraduate students in introductory ecology classes will have a manageable soils textbook. However, this book's information is for all ages. See it as an appetizer, a teaser, or an advertisement for soils. Everyone from the young naturalist to the home gardener can find something of interest in these pages. This book is your gateway to soils, soil science, and the world you tread upon every day.

Studying soils makes you better able to understand the world around you; soil science is truly an applied science. You will see how an understanding of biology, chemistry, physics, and ecology will open your eyes to the complexity of the world underfoot. You will learn the language that soil scientists use to communicate with each other. You will see that soils are dynamic and constantly changing. When left alone soils are resilient, but as we use them for our own purposes they can be fragile. It is in our own best interest and the interest of our human civilization to preserve them. Soils are classified in ways to help us make wise land use decisions as our population grows. Soils feed the world and will continue to do so. They dictated the rise and fall of ancient civilizations; have figured in art from the classics to the post-modernist, in warfare, and in literature. But soils are more than that; they are the foundation of life as we know it on Earth.

CO-EDITORS

David L. Lindbo, Ph.D., CPSS, NCLSS
North Carolina State University, Raleigh, NC

Deb A. Kozlowski, B.S.F., M.S.
Woods Charter School, Chapel Hill, NC

Clay Robinson, Ph.D., CPSS, PG
Stetson Engineers Inc., Albuquerque, NM

ACKNOWLEDGMENTS

Soil Science Society of America (SSSA) wishes to acknowledge and thank the Bureau of Land Management for funding and support of the development of this publication as part of their outreach activities to increase public awareness of our soil resources and promote responsible practices that reduce impacts to natural resources resulting from use of public lands.

SSSA is an international scientific society that fosters the transfer of knowledge and practices to sustain global soils. SSSA is the professional home for 6,000+ members dedicated to advancing the field of soil science. It provides information about soils in relation to environmental quality, ecosystem sustainability, bioremediation, waste management and recycling, crop production, and wise land use. A common thread across the programs and services of SSSA is the dissemination and transfer of scientific knowledge to advance the profession.

SSSA is also focused on outreach to students, teachers, and the public to tell the story of soil and build greater awareness of the value of soil to life. Visit our outreach sites:

www.soils4teachers.org
www.soils4students.org
www.iheartsoil.org

SSSA K-12 COMMITTEE

Chair: Clay Robinson; Members: J. Adewale Adewunmi, Suzanne Belflower, Meghan Buckley, Dedrick Davis, Jacob Elder, Sherry Fulk-Bringman, John Havlin, Margaret Holzer, Julie Howe, Amanda Mae Liesch, David Lindbo, Thomas Loynachan, Bradley Miller, Jarrod Miller, Bianca Moebius-Clune, Peter Narby, Robert Pesek, Susan Schultz, Melanie Szulczewski, Wendy Weitkamp;
Ex Officio Members: Susan Chapman, Emily Fuger

AUTHOR BIOGRAPHIES

WALE ADEWUNMI

J. Adewale Adewunmi is an Environmental Soil Scientist who does soil business in a non-traditional soil environment at a large regional wastewater facility in New Jersey. He manages land-application of enormous quantities of biosolids, conducts research and development associated with water quality issues and biosolids. Not knowing all the alternatives out there, he had hoped to work in soil science research and teaching, but stumbled into an attractive opportunity to solve real problems in this industry. Though he works among engineers, he brings the scientific method to solving process questions and address critical water quality issues. Since soil is fragile and has a limit to which it can be stretched before being contaminated, he learns how to balance industry needs with what nature can cope with in a scientifically sound and defensible manner.

Growing up, he has always thought of soil as "dirt," something for agriculture, until his first class in Soil Science where he heard terms like, *soil temperature*, *soil pH*, *soil water*, *acidity and liming*, etc. He was shocked to hear real science terms used together with soil, so he decided to find out more and follow wherever it led. He is always fascinated with fresh road cuts and deep construction sites where he can observe soil deeper. Adewunmi got his degrees in Agronomy, Soil Science, and Environmental Chemistry from Lincoln University, University of Missouri-Columbia, and Rutgers, The State University of New Jersey, respectively. He is a member of the Soil Science Society of America, American Society of Agronomy, Water Environment Federation, New Jersey Water Environment Association, and the International Union of Soil Science.

SUSAN CHAPMAN

Susan Chapman is the Director of Member Services for the Soil Science Society of America, where she is responsible for ensuring all of the members—from students to professionals—have the tools to help them build successful careers. As part of her job, she works with a great group of members on projects to help students and teachers learn more about the exciting world of soils and its importance to life on our planet.

Her career path working for associations for over 20 years, is not traditional, yet brought her to the Soil Science Society of America. She received her B.B.A. in Marketing and Management from the University of Wisconsin–Madison.

When not working with members, her favorite activities are triathlon (including Ironman), traveling, and reading a good book—like *Know Soil, Know Life*!

JOHN HAVLIN

A career in soil science does not mean "farming." John Havlin grew up in a large city—Chicago, Illinois. His interest in soils began while attending Illinois State University, where he was looking at career options for an undergraduate degree in chemistry. His undergraduate advisor suggested taking an introductory course in soils, and the rest is history. Following graduate degrees in agronomy (study of soil and plant sciences) at Colorado State University, he worked as a researcher and educator at the University of Nebraska, Kansas State University, and currently at North Carolina State University.

A career in soil science is fascinating and thoroughly enjoyable because you are working with one of the most complex systems on earth—soil. The activities you are involved in can be so diverse that you have no choice but to stay passionately engaged and stimulated to learn more. He has the opportunity to work with students at all levels, scientists from many disciplines, farmers, consultants, homeowners, and kids. During his career he has worked on numerous food and feed crops, including wheat, corn, soybeans, alfalfa, and many others. He has also worked with landscape plants, fruits, and vegetables, including grapes for making wine. Some days he is in a classroom teaching students, while other days he is working in the field or in a laboratory. He has traveled throughout the world working helping to protect and improve the soils ability to produce more food and to protect the quality of our water and air.

RICH HAYES

Rich Hayes is a licensed soil scientist who has been working professionally since 1985. He attended Davis and Elkins College and West Virginia University and has a B.S. degree. He is currently employed by the North Carolina Division of Water Quality Aquifer Protection Section, where he serves as a primary reviewer of soil reports and other related information associated with non-discharge permits.

Rich's professional experience also includes 18 years of working on soil surveys in Northampton, Halifax, Anson, Chatham, and Wake Counties in North Carolina. On the job activities included soil mapping and classification, as well as the collecting of soil data. As a project leader in Chatham and Wake Counties, Rich had the opportunity to set up seven new soil series.

Rich has always had a deep interest in conserving and protecting our natural resources, and since 2006 he has served as a Supervisor for the Chatham Soil and Water Conservation District. As a supervisor he also serves as the Area 3 representative to the North Carolina Association of Soil and Water Conservation Districts State Education Committee and as the Vice Chair of that committee. In 2009 Rich was the President of the Soil Science Society of North Carolina and an ex-officio member of the North Carolina Board for Licensing of Soil Scientists.

Rich has long been interested in environmental education and for many years has been active as a volunteer with the North Carolina Envirothon, where he has assisted with both Regional and State contests. He has been writing the State Contest soil tests since 2003 and has coordinated both the soils stations and the oral presentation component of the contest for the last few years. Currently Rich is the Vice Chair of the State Committee. Rich is also the principal author of the North Carolina Envirothon high school soils study guide.

DEB KOZLOWSKI

Deb Kozlowski was born in an industrial city in Connecticut, but knew from an early age that her future would lie in working with the natural world. "I've always been attracted to both science and art, and I find it odd when people think that's unusual—to me they are universally connected."

It was while attending the University of New Hampshire to study Forestry that she became interested in Soil Science. "When I first learned that I was required to take a course in Soils I was dreading it. Fortunately, an excellent professor had no problem convincing us of the importance of the subject, and showing us how fascinating a field it could be. I worked in his lab for the remainder of my years at UNH, and when I decided to attend graduate school at the University of Massachusetts, it was in Soil Morphology that I received my M.S." She spent the early part of her career working for an environmental engineering firm in western Massachusetts.

She never lost sight of the artistic side of soils. "When we moved to North Carolina I became involved in tile making at a small arts center in eastern NC. There was a limitless supply of clay to dig!" Today she is the Art teacher at Woods Charter School for grades K-5. "Kids have no prejudice when it comes to combining science and art—the messier the better." She still dons her lab coat as a special guest lecturer on any soils topic—from the chemistry of clay to experimenting to learn which soils are best to sprout beans in a paper cup.

She divides her year between Massachusetts and North Carolina, where she lives with her husband and two sons on a small farm south of Chapel Hill. They share many interesting dinnertime conversations, which may not be appropriate for polite company. "We all know where our food comes from," she says, "and we all know where it ends up."

MANDY LIESCH

Mandy Liesch is a graduate student in soil physics at North Carolina State University. She is originally from Northern Wisconsin, and she loves cheese and the Packers. She took agricultural and science classes in high school because she wanted to be a veterinarian. She actually avoided taking soils and crop related classes because only farmers needed to know that stuff. It was easy, plant seed, give it food, and wait. Later in life, she learned that there was so much more to food production and soil health than she thought. Even though she liked science, she preferred playing basketball and track, the saxophone, and studying social studies and politics, including how agriculture is important around the world. This interest in people caused her to major in International Studies at the University of Wisconsin–River Falls, with an agricultural development emphasis. She didn't even know that soils existed, but it was required in the curriculum. She fell in love with soils because of a charismatic soil science professor. She decided to minor in soils and went on to study soils during her Master's degree at Kansas State University.

Mandy loves teaching soils, especially to college students and to youth, and has worked with the local Girl Scout Councils developing environmental and soil science programs since her freshman year of college. These include camp gardens, a greenhouse made out of pop bottles, and various soil arts and craft projects. Soils are the coolest material on the planet, and she sets out to make sure that everyone knows that soils are responsible for the food that she loves. Mandy uses soils as an excuse to travel around the world and has been to four continents, tasting the different foods and observing how soils are unique. Mandy really loves incorporating soils into her jewelry designs, and using soils, especially clays to create paints and natural dyes. She also loves making cheese, kayaking, rock climbing, and doing dog agility with her lab mix, Clark.

DAVID LINDBO

Dave Lindbo is a professor of environmental soil science in the Department of Soil Science, North Carolina State University and lives in a log cabin near the Haw River with his wife (an artist and soil scientist) and two sons. He grew up near Boston, MA and was blessed with parents and grandparents who still had a strong connection to the land. His first real experience with soils (other than mud pies and excavating with a Tonka truck) was in elementary school when he began working with his grandfather in his extensive vegetable garden. "My grandfather wanted his grandchildren to understand where food came from and more importantly how to grow it without hurting the environment."

This connection to the land led him to majoring in environmental conservation and geology at the University of New Hampshire, where his interest in soil was rekindled by the unique teaching style of Dr. Nobel K. Peterson that made soils fun, engaging, and informative. "He showed students that life as we know it could not exist without soil, since our food comes from the soil as well as the materials for our homes, the fiber for our clothes, and a host of other products." He continued his studies earning an M.S. degrees in Soil Science (UNH) and Geology (UMass) and a Ph.D. in Soil Science (UMass).

His professional career has included work in water quality, research with USDA-ARS in Mississippi, teaching at UMass and for Mass DEP, and his current position at NCSU. "I feel I can continue to spread the word about how critical soils are to our everyday life." Part of his job is teaching K-12 and Envirothon students (including his sons) and teachers about the importance of soils—while having fun—whenever possible. "If you dig a soil pit

for someone they may be dirty for a day, but teach them to dig and they'll be filthy and fulfilled for life. Getting dirty is the best part of my job."

TOM LOYNACHAN

Tom Loynachan (pronounced "law-na-cun") is a professor of soil microbiology at Iowa State University. He grew up on a farm in south-central Iowa and had his hands in the good soils of Iowa early in life. In high school, Loynachan learned that Iowa soils are some of the finest in the world, and this made a lasting impression on him. At about the same time in high school, a gifted teacher stimulated his interests in biology. Early in college, Loynachan discovered that he enjoyed teaching and working with students. After graduating with his B.S. degree, Loynachan taught vocational agriculture at Grinnell, Iowa. Shortly thereafter, he was drafted into the army. After an honorable discharge, Loynachan continued his education earning M.S. (Iowa State University) and Ph.D. (North Carolina State University) degrees in Soil Science.

Loynachan finds life in the soil fascinating and enjoys sharing his excitement with others. He routinely visits high schools and discusses "life in the soil," and he teaches a college senior-level laboratory on soil biology. Also, Loynachan has created short movies that describe the various groups of organisms in soil and what they do. He feels one trait of an effective teacher is to pass the excitement of the subject on to his students. Loynachan emphasizes that much of life in the soil is still unknown, and there are great opportunities for future discoveries by the next generation. Loynachan hopes that some of his enthusiasm for the living component of soils will stimulate a future scientist, or perhaps provide others who are not destined to become scientists, with an appreciation for the complexity and diversity of the living soil. Loynachan believes that soils are unmistakably one of the world's greatest natural resources, and they must be maintained for all future generations.

DOUG MALO

Doug Malo is a Distinguished Professor of Pedology in the Plant Science Department at South Dakota State University, where he has been teaching and doing research for 36 years. He grew up on a small family farm in south-central Minnesota. While living on the farm, he learned to appreciate the role of soil health in maintaining soil productivity. On the farm they raised corn, soybeans, alfalfa, peas, sweet corn, and oats, along with beef cattle, chickens, and pigs. While working on his B.S. degree in agronomy and plant pathology, his interest in soils grew when he became involved in the Iowa State University soil judging team. Upon graduation, Doug decided to further his education and attended North Dakota State University, where he earned his M.S. and Ph.D. in soil genesis and classification.

His passion is to teach soils to inquisitive minds. One of his mottos is: "Students do not interrupt my work, they are my work." After almost four decades of soil genesis research and teaching soils classes, he still has enthusiasm for the subject matter, but his real joy in teaching is coaching the SDSU soil judging team. Students from many different majors have been members of his teams and have become successful professionals in many agricultural and natural resource fields. The goal in soil judging is to instill the love of learning about soil, so that his students become better stewards of our soil resources and to protect those resources for future generations.

BIANCA MOEBIUS-CLUNE

Bianca Moebius-Clune is an Extension Associate in the Department of Crop and Soil Sciences at Cornell University. She grew up in Germany, surrounded by farms, and learned to grow fruits and vegetables in her backyard early in life. When she moved to Durham, NH with her family, she first found out how beautiful soils can be while digging holes in the forest, discovering amazing horizons of black, white and red soils! (It was much later she realized that she had happened upon a Spodosol!) Bianca always enjoyed outdoor activities, particularly playing with soil and plants. She developed a passion for the environment throughout her years in school, but didn't discover soil science until she started college.

While working on her B.S. in soil science at the University of New Hampshire, she studied abroad in Madagascar for six very influential months. The people she lived with in small Malagasy farming villages grew all their own food—those who had damaged their soils more often went hungry, those who managed them well had plenty of food and could afford to send their children to good schools. She recognized just how much we all depend on our soils—for food, clean air, clean water, clothing, housing, recreation. She often likes to challenge students to discover how everything in our life is directly or indirectly linked back to soil. Inspired, Bianca went on to Cornell University to earn an M.S. and Ph.D. in soil science. She did much of her work on developing the Cornell Soil Health Test in the northeastern United States, but also got to travel to Kenya. These days her position allows her to do many of the things she loves: teaching students and farmers about sustainable soil management, traveling, researching ways we can manage our soils better, and she is still digging holes in soil, too. She notes that soil science is not only fun, but it opens many doors to opportunities for making the world a better place.

CLAY ROBINSON

Clay Robinson has been a professor of soil science and is now a consultant, always striving to help people be better stewards of their soil and the environment. He grew up in a small Texas Panhandle town, helping his dad work in a small vegetable garden and making compost to enrich the soil. Clay is a great name for a soil scientist, although it had drawbacks—his uncles called him "Mud." He started college as a computer science major, but realized he needed a job where he could be outside. Five majors later, he finished with a B.S. in agriculture, followed by an M.S. in plant science from West Texas State University, and finishing with a Ph.D. in Soil Science from Iowa State University.

He has a passion for sharing knowledge, which led him to be a college professor, and later to become "Dr. Dirt," taking a message to kids about the importance of taking care of the limited soil so that everyone does not end up hungry, naked, and homeless.

Clay enjoys acting, singing, playing the guitar, cycling, hiking, and photographing landscapes, sunrises, and sunsets. He even uses soil as a natural pigment to paint landscapes.

MELANIE SZULCZEWSKI

Melanie Szulczewski grew up in a city and understands that most people do not give much thought to the soil—she was one of those people! She believes, however, that once someone starts to reflect on all that soils do for us, even in the city, they will develop an incredible appreciation for soil science. Szulczewski majored in chemistry and French literature at Cornell University. Shortly before graduation, she realized that she wanted to get out of the laboratory and apply chemical concepts to the world outside. She discovered the field of soil science by accident and fell in love with the amazing world beneath our feet. She earned master's and doctorate degrees in soil science at the University of Wisconsin–Madison, specializing in soil chemistry and degraded environments. Since then Szulczewski has worked on projects in the Everglades, sub-Saharan Africa, and the Chesapeake Bay watershed. She is currently an assistant professor of environmental science at the University of Mary Washington in Fredericksburg, VA.

As an environmental scientist, Szulczewski believes it is important to see the connections between all the diverse disciplines required to create a sustainable, healthy world. The state and treatment of our soil affects many different environmental problems: food production, pollution issues, and biodiversity, just to name a few. She believes learning more about the soil will create more knowledgeable citizens no matter what their career paths may be.

xviii

CHAPTER 1

KNOW SOIL, KNOW LIFE

The world around us is teaming with life. We see it every day. We study it in school so we can fully appreciate and know something about the natural living world around us. But how deeply do we really know life? We know that it is necessary in all its complexities and diversity for our well being. We know that life on Earth makes us unique in the solar system. But rarely do we ponder the fundamental connection between soil and life. Do we ever consider that with *no soil* there would be *no life*?

This book will introduce you to an amazing world—the world beneath your feet. Soil is the foundation our natural living world depends on, the often-unappreciated substance of life, the *dynamic* material that civilization is built on, the *critical zone* of the earth. Once you *know soil* in its complexity and beauty, you will *know life* with broader horizons. You will see that "heaven is under your feet as well as over your heads," in the words of Henry David Thoreau. Soil is not dirt. Soil is life!

This chapter highlights the importance of soil to our everyday lives and introduces some basic facts about soil that will be explored in depth in this book.

SOIL'S IMPORTANCE TO PEOPLE

As you consider the world around you, it comes as little surprise that everything depends on a few basic things: food, a place to live, and water. In the bigger picture, energy is critical, the ultimate source of which is the sun.

People have a few other particular needs. We need air for oxygen, as well as fiber for clothing. Upon further consideration it may come as a surprise that much of what we depend on—food, water, fiber, shelter—are all related to a single, often overlooked item. This is soil! Soil (the *pedosphere*) represents the critical zone of the earth where life (the *biosphere*), water (the *hydrosphere*), minerals (the *lithosphere*), and air (the *atmosphere*) intersect and interact (see figure 1–1). We are reminded that to know soil is to know life, and that with no soil there is no life.

A close look at what you ate for breakfast illustrates this point. You may have had cereal, milk, and orange juice, or maybe toast. Where did each of these items come from? The wheat in your cereal or flour in your toast started out as plant

CHAPTER AUTHORS

DAVID LINDBO
JOHN HAVLIN
DEB KOZLOWSKI
CLAY ROBINSON

seeds. The milk came from a cow that ate grass. The juice came from oranges that grew on a tree. If you ate sausage or bacon remember that meat comes from animals that are fed grain and forage, which also come from plants. So all your breakfast foods, and food in general, can be traced to plants. But think about where plants come from: plants grow in soil! The soil provides water as well as the nutrients plants need to produce the food we eat. When you eat, you are eating soil, although several steps removed. We rely on plants to supply us with the food we need to survive. Without soil to produce the plants we would have no grains, no bread, no cereal, no milk, no meat, no fruit, no pizza, and therefore would not be able to survive (figure 1–2).

Next, consider your clothes. Your t-shirt and jeans are probably made of cotton—a plant that depends on soil. Other natural

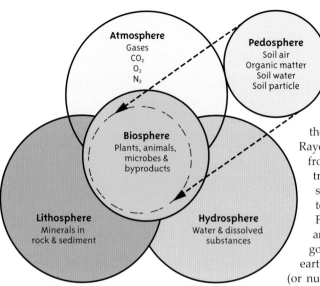

Figure 1–1. The intersecting spheres on Earth: rock, life, air, water, and soil.

fibers like wool, linen, and silk are also directly related to plants—wool comes from animals that eat plants, linen from flax plants, and silk from silk worms that eat plants. Even many synthetic fibers can be related to soil. Rayon is a synthetic fiber made from cellulose, which come from trees and other plants. Other synthetic fibers such as polyester are derived from fossil fuels. Fossil fuels are ancient plant and animal remains that have undergone extreme changes deep in the earth but nonetheless needed soil (or nutrients from soils and weathering) when they were living. Through a few steps we can indeed link fossil fuels to soil.

Finally, think about the home you live in. In general, houses are made from lumber and bricks. It is easy to see that bricks are connected to the soil, since they are made from clay and sand. Lumber takes a few more steps to connect to soil: lumber is wood; wood comes from trees; and trees, like other plants, need soil to survive as we saw above. Very simply, your house as you know it would not exist without soil.

SOIL'S IMPORTANCE IN THE ENVIRONMENT

We've seen that soil is critical to our food, fiber, and shelter. It also plays an important role in the cycling of another essential item: water. The amount of water on Earth is constant--there is no loss or gain of it. And only a small portion of

Figure 1–2. (a) Everything in this shopping cart comes from soil. (b) Sod house, stick built house, green earth house—all from soil. (c) Clothed in mud or elegant dress—all from soil. Read the label on your shirt and think about how the textile comes from soil. Cotton? Silk? Polyester? Rayon? Wool?

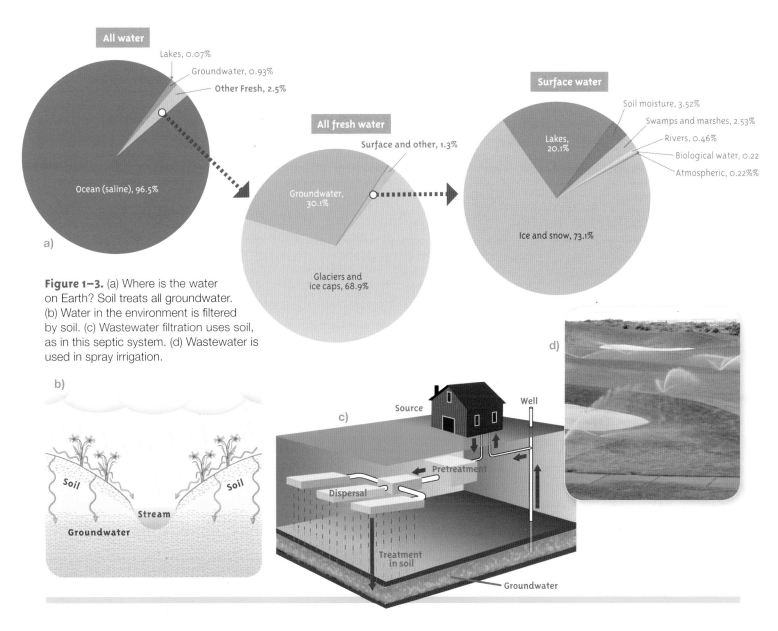

Figure 1–3. (a) Where is the water on Earth? Soil treats all groundwater. (b) Water in the environment is filtered by soil. (c) Wastewater filtration uses soil, as in this septic system. (d) Wastewater is used in spray irrigation.

that finite amount (approximately 1%) of water is usable for drinking (figure 1–3). How does this water remain clean enough for us to drink? You might think it's because we treat it before we drink it. While this is true in many urban or suburban areas, many people worldwide get their drinking water from groundwater. This groundwater is often not treated before people drink it. Instead, it is treated by the soil. As water infiltrates and percolates through soil, the soil's chemical, biological, and physical properties clean the water by removing contaminants. This means that soil is perhaps the largest single water (and wastewater) treatment plant in the world. Soil helps keep water clean by filtering it. We'll learn more about the role of soil in water cycling in chapter 2.

We've seen how soils are essential to plants, by providing many of the nutrients that plants need. Where the supply of nutrients is too low for *crops* to grow we add nutrients to ensure growth and a reliable food supply. Soils store the nutrients until the plant needs them. Soil acts as a nutrient reservoir for plant growth and survival. Soil also provides critical support for plant roots, preventing the plant from falling down or washing away. Soil also acts as a sponge to hold water. Plant roots take up water, allowing the plant to grow and photosynthesize.

In addition to its environmental roles relating to food and water, soil also plays a role in construction. Properties of soil are important to consider when constructing roads and buildings that will last. In particular some soils will shrink and swell because of changes in the amount of water present (figure 1–4). Engineers must identify these and other soil properties to properly design structures (figure 1–4).

Beyond our food, fiber, and shelter, soils are intimately involved in construction and water treatment. Now that we understand how important soil is to our lives, we can consider how much of this precious resource we have.

HOW MUCH SOIL IS ON EARTH?

Surprisingly, there is not that much soil on Earth, yet it is one of the most important natural resources. **As the world population increases, the finite soil resource must provide enough food, fiber, and shelter for the world.**

Relatively speaking, how much productive soil do we have?

The earth has approximately 149 million square kilometers (58 million square miles) of land area (figure 1–5). Of this, deserts and ice sheets account for about 31% and forests another 31%. The remaining 38% is considered agricultural land, but 26% is in permanent pasture, used only to produce feed for livestock, such as cattle, sheep, and goats. Only about 12% of Earth's land surface is used to produce food and fiber (cotton) for human consumption. Of that, just over 1% is perennial cropland primarily used for orchards and vineyards. The remaining 11% is considered *arable* land, which is capable of sustaining annual crops. In the United States, about 23% of the land is in deserts or mountain ranges, 33% is in forests, 26% is rangeland used for grazing, and 18% of the land is used for producing crops.

Since 1961 the global amount of land in annual crops has varied between 9.5% and 11%. In the same period, the world

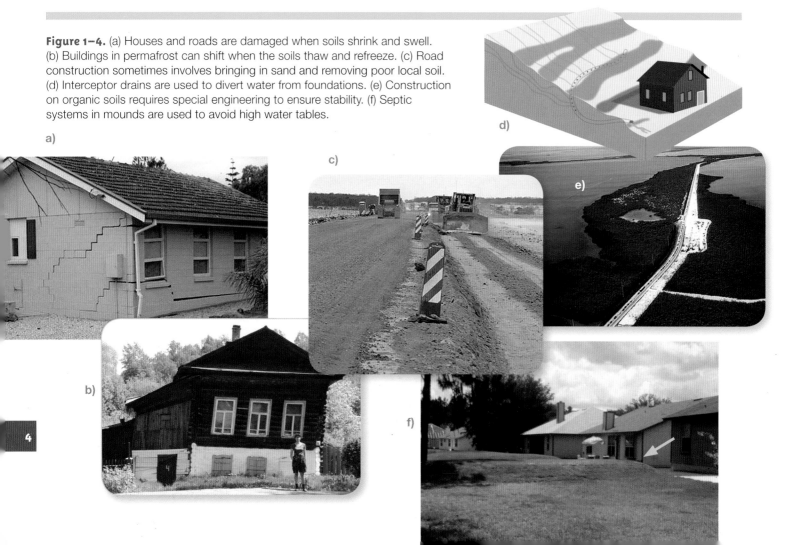

Figure 1–4. (a) Houses and roads are damaged when soils shrink and swell. (b) Buildings in permafrost can shift when the soils thaw and refreeze. (c) Road construction sometimes involves bringing in sand and removing poor local soil. (d) Interceptor drains are used to divert water from foundations. (e) Construction on organic soils requires special engineering to ensure stability. (f) Septic systems in mounds are used to avoid high water tables.

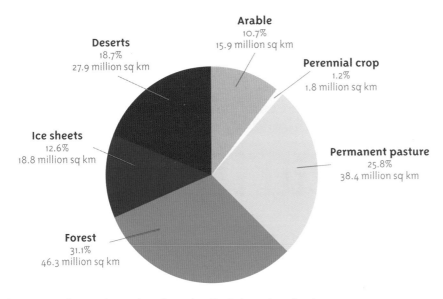

Figure 1–5. Comparison of total area on Earth in various land uses.

average of arable land per person has decreased from 0.37 hectares per person in 1961 to 0.20 hectares in 2012. (Note that a hectare, or 10,000 square meters, is approximately 2.5 acres, and one acre is 43,560 square feet). Nor is the arable land evenly distributed. In East Asia and the Pacific, less than 0.10 hectare per person is available, while Africa has about 0.20 hectares and North America has about 0.61 hectares. As the human population grows, arable land comes under increasing pressure to produce more food per hectare. Some countries don't have the economic strength to buy fertilizers, better seed, and other inputs required to increase yields, or they lack the means or water supply to irrigate, so they look for more land to produce crops. But converting other land into food production poses problems.

Many of the world's forests are in areas that are too cold to produce food crops. Others are in steep areas or shallow soils over bedrock. Removing the trees from these soils leads to rapid *erosion* and loss of productivity. Other forests are in high rainfall regions with acid soils requiring many amendments and careful management to maintain productivity. The capital to purchase the inputs and management expertise in some of these regions is limited.

Deserts are fragile ecosystems that receive too little precipitation to grow crops. Grazing lands (permanent pasture) are often in semiarid regions, and are highly susceptible to drought. Plowing such lands to produce food already led to one Dust Bowl in North America in the 1930s (see chapters 6 and 8) and is having similar impacts in Asia and Africa now. Some soils in deserts and semiarid regions have so much salt in them that plant growth is limited.

Irrigation can increase crop yields and decrease drought risk for crop production in arid and semiarid regions. However, water supplies are becoming more limiting in both quantity and quality. Overall, less than 1% of the arable land in the world is irrigated. About 5.5% of arable land in the United States is irrigated, while more than 50% of the arable land in Pakistan, South Korea, and Bangladesh is irrigated. Worldwide approximately 40% of all food crops are irrigated.

Medieval alchemists considered there to be four elements: earth, air, fire, and water. You can think of these as soil, air, sunlight, and water (figure 1–6), the four items critical to life on Earth. We know we cannot spin gold out of the four medieval elements, but we do need these four things to support something more valuable: all life as we know it.

Earth as an Apple

Grasping land in terms of acres, hectares, square miles, or square kilometers may be hard to fully comprehend. Another way to consider the relative amount of arable land is to imagine Earth as an apple. Cut the apple into four equal parts and discard three parts (75%)—that represents the oceans of the world. The fourth part (25%) represents the land area. Cut the "land area" part in half, leaving two parts each equal to one eighth (12.5%) of the total apple. Discard one of the parts as it represents the deserts, wetlands, and polar areas where people do not live.

The remaining part (12.5%) represents the areas where people live, but not all of it is used for producing food. Cut this remaining part (12.5%) into four pieces. Each remaining slice represents about one thirty-second (1/32, or 3.125%) of the original apple. Three of these one thirty-second sections represent areas of the world that are too rocky, too wet, too hot, or where soils are too poor for growing food, as well as urban areas. Thus, only one thin slice of apple (3.125%) is suitable for food production. But also consider this: soil is present only at the very surface of the earth, so peel that last slice. The thin apple peel represents the soil where all our food, fiber, and shelter come from. Each year more and more of that thin sliver is lost from production, yet the human population continues to grow (see the end of this chapter and chapter 6). This makes it more and more important to take proper care of the soil we have.

Figure 1–6. A modern view of the alchemists' four elements we cannot live without.

WHAT IS SOIL?

Realizing the importance of soil and how little there is, we need to briefly make sure we all know what *soil* is (figure 1–7). First and foremost, **soil is not dirt.** Dirt is the stuff under your fingernails; it is what you sweep up off the floor; it is unwanted and unnecessary. Soil, on the other hand, is essential for life, so soil is not dirt! Dirt may be soil out of place, just as a weed is a plant out of place. For example, a rose in a cornfield, while beautiful, is a weed; a corn plant in a rose garden is also a weed. So when you track mud (wet soil) inside, you are putting the soil in a place it is not wanted. At that point it becomes dirt.

If soil is not dirt what is it? There are several definitions…. Perhaps the simplest is that soil is a living, *dynamic* resource at the surface of the earth. To expand that definition, soil is

a natural, three-dimensional body at the Earth's surface. It is capable of supporting plants and has properties resulting from the effects of climate and living matter acting on earthy parent material, as conditioned by relief and by the passage of time.

Worldwide tens of thousands of different soils occur on every continent and virtually anywhere plant life can set roots. An understanding of the environment requires an understanding of soil—what it is, how it is formed, what it is made of, and how it is used. Soil serves as a repository of many geological and climatic events that have occurred in its location. It is a window to the past, but it can also serve as a view of the future as its properties relate to how we can and should manage this finite resource.

Now consider a handful of soil. At first it may seem lifeless and solid, but in reality soils are teaming with life and contain pockets of air and water. There are four components to every soil: minerals, organic matter (living and dead), water, and air. The minerals and organic matter make up the solid phase. The water and air make up the pore space. A typical handful of soil contains 50% pore space, 45–50% minerals and 0–5% organic matter (figure 1–8). These components will be discussed in detail in the next chapter. Soil as a whole is affected by four basic processes, discussed next.

BASIC SOIL PROCESSES

That same handful of soil is dynamic and responds to its environment. A great number of processes take place in the soil, but they can be grouped together under four major categories: *additions, losses, transformations,* and *translocations* (figure 1–9). Each of these is briefly defined in the following paragraphs.

Figure 1–7. What is the difference between soil and non-soil? Non-soil includes dirt on the floor, sand dunes in deep desert, barren rock areas, and surface materials at extreme elevations. Think about where the soil is in the world around you—in an agricultural field, lawns, parks, and gardens in urban areas, as in New York's Central Park. There is even soil in the Artic and Antarctic desert, subaqueous soil with eel grass and seaweed, and where trees appear to grow straight out of rock.

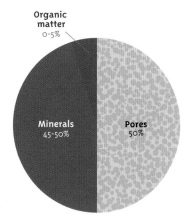

Figure 1–8. Percentages of each component of soil: mineral, pore space (filled with air and/or water), and organic matter.

ADDITIONS: Rain adds water. Dust adds minerals. Animal wastes, leaf litter, and dead roots add organic matter and nutrients. Humans add fertilizers.

LOSSES: Water evaporates into the air. Nutrients are taken up by plants. Soil particles wash away in a storm. Organic matter may decompose into carbon dioxide. Minerals and nutrients leach into groundwater.

TRANSLOCATIONS (movement within the soil): Gravity pulls water down from top to bottom. Evaporating water draws minerals up from bottom to top. Organisms carry materials every which way!

TRANSFORMATIONS (one component changes into another): Dead leaves decompose into humus. Hard rock weathers into soft clay. Oxygen reacts with iron, "rusting" the soil to a reddish color.

Figure 1–9. The four essential processes of soil: additions, losses, translocations, and transformations.

Additions are easy to understand. They consist of materials deposited on the soil from above, as well as materials moved in with groundwater, such as salts. Obvious examples are additions of leaf litter as trees shed their leaves, or additions of organic material as plants and plant roots die. Also obvious are additions of mineral material from flooding, landslides, and other geologic events. Perhaps not so obvious is the nearly constant addition of atmospheric dust to the soil surface. Some of this dust can travel long distances and is important to the overall fertility of a region. Rainfall is also an addition.

Losses are also rather obvious. Erosion is a major form of soil loss. Loss can also occur as nutrients are taken up by plants,

Across the Seas

One of Africa's biggest "exports" is soil from the Sahara Desert. Whipped up by winds and carried into the atmosphere, fine particles of desert soil can travel across the entire Atlantic Ocean. The dust sometimes causes hazy skies in the southeastern United States and is even credited with adding needed nutrients to the soils of the Amazon rainforests.

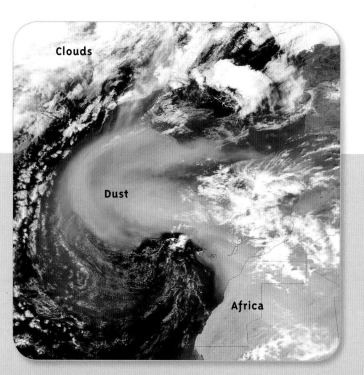

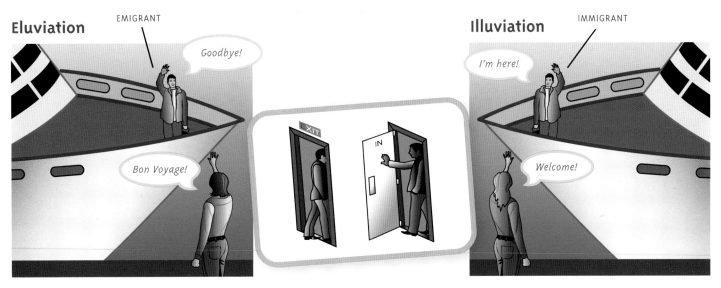

Figure 1–10. Illuviation and eluviation can be understood with these metaphors—the first letter of these words will help you remember! **E**luviation is like **e**migration, leaving your country or origin to go to a new country. **Il**luviation is like **i**mmigration into a new country. **E**xit is leaving a room, as **e**luviation is material exiting the horizon. **In** is moving into a new room, as **il**luviation is material moving **in**to a zone below.

and plants are harvested and removed. As minerals and nutrients move through soil into groundwater or out of the plants' rooting zone, this too is considered a loss.

Translocations are similar to losses in that they involve movement of materials. Translocation differs in that the material is not removed from the soil; instead, it moves from one location to another. This internal movement is referred to as *illuviation* and *eluviation*. Eluviation removes material from a zone. Illuviation moves material into a zone. In other words, eluviation is material exiting, while illuviation is material entering (figure 1–10).

Understanding transformations takes a little more thought. Soils are dynamic—that means they are constantly changing, and biological, chemical, and physical transformations are part of this. For example, leaf litter falling on soil eventually decomposes. This decomposition is a transformation process. Likewise as rocks weather to soil, this too is a transformation process. The initial minerals in the rock are transformed to clays in the soils over time. One mineral can be transformed to another without additions of materials.

The cumulative processes that act on a soil depend on a range of environmental factors, which in turn shape the world's different biomes, discussed next.

SOILS AND BIOMES

Ecologists group large geographic regions with similar environments and distinctive plant and animal communities into biomes (figure 1–11). The major terrestrial biomes include savanna and temperate grasslands, tropical and temperate rainforests, boreal and temperate forests, arctic and alpine tundra, deserts, shrublands, and wetlands. Each biome comprises several ecosystems. For example, temperate grasslands may be short, mixed, or tallgrass prairies.

The environmental factors influencing biomes include latitude, the general climate and topography of the region, and soil. Soil is the foundation of every terrestrial ecosystem.

Each biome has soils with characteristics unique to it. Soil is more than just weathered mineral particles of different sizes. Even a handful of soil is home to more organisms than people living on Earth, as we will learn in chapter 3. Most of these organisms are too small to be seen without a microscope. Microbes consume dead plant and animal tissues to get energy and nutrients. When more nutrients are available than the microbes need, the surplus nutrients are released into the soil. These nutrients are available to support new plant growth, which in turn support animals.

Many plants and animals either prefer, or are adapted to, specific types of soils within a biome. These connections among soils and biomes run both ways; the soil influences the biome, and the biome affects the soil found in it. These relations among soils and biomes will be explored in greater detail in chapter 7.

POPULATION GROWTH AND SOIL

People have exerted a huge influence on biomes over time. We are converting natural biomes into urban lands and cultivated lands (both forests and farms) (figure 1–12). As population growth continues we use more of our soil resources. Human population growth since 1950 (figure 1–13) has caused

Figure 1–11. Earth's biomes include savanna and temperate grasslands (prairie), temperate and boreal forests, tropical and temperate rainforests, deserts, shrublands, alpine and arctic tundra, shrublands and wetlands. These are merely examples. Can you think of a forest that looks different than this one? How about a different desert?

rapidly growing demands for food, water, timber, fiber, and fuel. These demands are affecting ecosystems more extensively than in any comparable time in human history. Approximately 12% (1.55 billion hectares) of total world land area and 32% of agricultural (arable) land (4.93 billion hectares) is currently cropland (table 1–1). The remaining 3.38 billion hectares of agricultural land, primarily (90%) in Latin America and sub-Saharan Africa, is in forests, permanent pasture, and other non-crop uses. To expand cropland into these 3.38 billion hectares, farmers face large costs because of the poor soil fertility, shallow soils, low rainfall, and other limitations. Furthermore, loss of biodiversity, soil, and other factors affect ecosystem functions, such as maintaining water and air quality.

Since the early 1960s, the world annual population growth rate has dropped to slightly over 1% (figure 1–13). While annual growth rates in all nations are projected to decrease by 2050, rates in developing nations will still be approximately sixfold higher (0.5%) than those in industrialized nations (−0.1%) (table 1–2). In developing countries, most of the decline in population growth rates is projected to result from improvements in education, economic development, and agricultural productivity, primarily in South and East Asia, where nearly 50% of the world's population resides. In sub-Saharan Africa where poverty, suppressed economic opportunity, and low agricultural productivity persist, absolute populations are projected to more

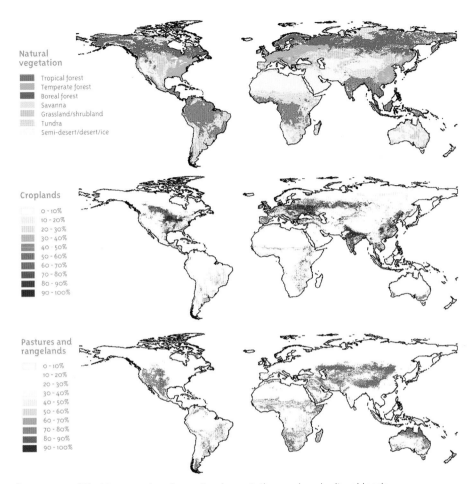

Figure 1–12. World maps showing natural vegetation and agricultural lands.

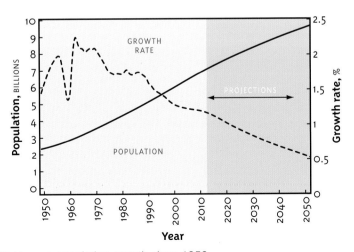

Figure 1–13. Human population growth since 1950.

Table 1–1. Approximate world land area used for food production.

	Million ha
World	13009
Agricultural land	4932
Cropland	1554
Cereals	700
Oil crops	251
Pulses	73
Roots and tubers	55
Vegetables and melons	52
Fruits	47
Fiber crops	36
Tree nuts	8

than double to 1.8 billion in 2050. It is interesting to note that population in developed nations will decrease from 32% of total world population in 1950 to about 13% by 2050 (calculated from data in table 1–2).

Total agricultural land has been relatively constant since 1990, whereas arable and permanent cropland, has increased slightly, likely into less productive areas (figure 1–14). Per capita cropland use decreased nearly 50% from 1960 to 2008 (figure 1–15). By 2050, world cropland per capita is further projected to decrease by another 30%, assuming the amount of cropland stays constant.

The other trend to note is that the number of people living in urban areas is increasing; for the first time in history, in 2008 more than 50% of the world's population lived in urban areas. By 2050 more than 60%, or nearly 6 billion people, are projected to live in urban areas. Therefore, the impact on cropland is somewhat lessened by the growth of urban areas. Urban population growth, however, commonly occurs on highly productive lands, and urban expansion in developing countries decreases cropland by 0.5 million hectares per year.

Table 1–2. Differences in population growth and annual population growth rate between developed and developing countries.

	Population						Average annual growth rate				
	1950	1970	1990	2010	2030	2050	1950–1970	1970–1990	1990–2010	2010–2030	2030–2050
	billions						%				
World	2.54	3.70	5.29	6.91	8.32	9.19	1.99	1.74	1.21	0.82	0.41
Developed countries	0.81	1.01	1.15	1.23	1.26	1.25	0.96	0.59	0.32	0.06	−0.09
Developing countries	1.72	2.69	4.14	5.68	7.06	7.94	2.41	2.08	1.41	0.96	0.49

Table 1–3. Distribution of global lands and population in land quality classes. The higher the Land Quality Class number, the greater the limitations for sustainable grain production. In other words these classes are generally less productive and more easily degraded lands.

	Land quality class†	Land Area		Population‡	
		million ha	%	millions	%
Decreasing land productivity ↓	I	409	3.1	337	5.9
	II	653	5.0	789	13.8
	III	589	4.5	266	4.6
	IV	511	3.9	654	11.4
	V	2135	16.4	1651	28.8
	VI	1722	13.2	675	11.8
	VII	1165	8.9	639	11.3
	VIII	3996	28.3	103	1.8
Total		13058	100	5739	100

† Includes risk for sustainable grain crop production: Class I <20%, Class II and III 20–40%, Class IV–VI 40–60%, Class VII 60–80%, and Class VII and VIII >80% risk.

‡ Population only between 72°N and 57°S latitudes.

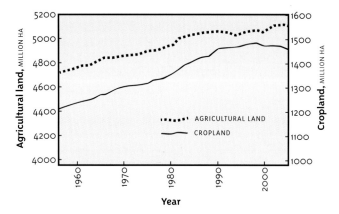

Figure 1–14. Change in world agricultural land and cropland since 1960. Cropland represents arable and permanent cropland.

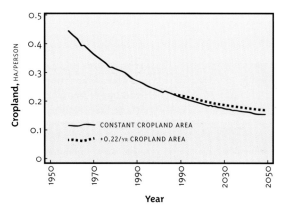

Figure 1–15. Historical and projected per capita world cropland use. Projections to 2050 are based on no expansion of current cropland area of 1.55 billion ha (solid line) or the average annual rate of cropland increase from 1996–2007 of ~3.38 million ha/yr, or 0.22%/yr (dashed line).

The growing population's demand for food, fiber, and fuel will obviously not be met by expanding cropland area, since marginal, low producing land is all that is available for conversion. Soil scientists classify land according to its suitability for cropping (see chapter 5). Unfortunately, only 13% of the global land surface is considered prime cropland (Class I) or land with few problems limiting sustainable grain production (Class II & III) (Table 1–3). Approximately 76% of the global population resides on the least productive lands (Class IV–VIII). This may seem alarming; however, with equitable food import–export systems, crop production from the most productive lands can and has met global food and other resource demands. Although many factors limit crop productivity, soil moisture and temperature stresses are the main ones. Cropland productivity can be limited by numerous climate and soil properties that will be discussed in subsequent chapters.

Since we can't increase the amount of land suitable for crops, future increases in food production will have to come from increasing how much food can be grown on the existing cropland. We will have to learn to sustain and enhance soil productivity, especially in developing countries. We will also have to reverse current trends in soil degradation resulting from improper soil management that are threatening our capacity to meet future food demands. Removal of a significant proportion of field crop residues for fuel production (biofuels and ethanol) also may further accelerate soil degradation.

SUMMARY

As our population has grown, people around the globe have implemented ways to conserve the soil we depend on. Despite these efforts, worldwide, we are still losing a staggering 26.4 billion tons of soil every year, a far greater amount than is being replenished.

This introduction should help you understand why that loss matters, and the importance of soil to everyday life and to our continued existence on the planet. In subsequent chapters we shall expand on the dynamics and subtleties of this often overlooked yet critically important resource to allow you to understand and appreciate that soil is not dirt and that soil sustains life. We hope you will come to know soil and thus know life.

CREDITS

1–2b, earth home, Peter Vetsch; 1–2c, Asaro mud man, Michael Johnson, MLJohnson.org; 1–3c, irrigation, GCSA, Scott Hollister; 1–4a, house on shrink–swell soil, Geoconsult Geotechnical Consultants; 1–4b, house on permafrost, Adam Jones, Ph.D./Flickr; 1–4c, road construction, North Dakota Parks and Recreation Department; 1–4d, construction on organic soil, Hayward Baker Geotechnical Construction; 1–4f, septic mound, Frank Watts; 1–7, lichen on tundra, Hugh Ducklow, trees in scree, David Lindbo; Sahara sidebar, NASA/SEAWIFS; 1–12, Science. Chapter opener image, iStock. Table 1–1 and Table 1–2 data from UN-FAO (2007); Table 1–3 data from Beinroth, Eswaran, and Reich (2001).

Copyright © 2012. Soil Science Society of America, 5585 Guilford Rd., Madison, WI 53711-5801, USA. Know Soil, Know Life. David Lindbo, Deb A. Kozlowski, and Clay Robinson, Editors
doi:10.2136/2012.knowsoil.c1

GLOSSARY

ADDITIONS The process of depositing new components/materials into the soil materials, either from above, or moved in with groundwater.

ARABLE Land that is capable of sustaining crops, where production is practical and economically feasible.

ATMOSPHERE The envelope of gases surrounding the earth (the air).

BIOME A large geographic region with similar environment and distinctive a plant and animal community.

BIOSPHERE The regions of the surface and atmosphere of the earth or other planet occupied by living organisms (the biota, or living organisms).

CRITICAL ZONE Soil or pedosphere—the zone or area at the surface of the earth where life (the biosphere), water (the hydrosphere), minerals (the lithosphere), and air (the atmosphere) intersect and interact.

CROP A plant used for human purposes, such as food, fiber, construction material or fuel.

DYNAMIC A system that is active, and constantly changing due to many interactions among components.

ELUVIATION The translocation (removal) of soil material (either suspended or dissolved in water filling soil pores) from a layer (soil horizon) to a deeper horizon. Usually, the total loss of material from the soil profile to groundwater is called "leaching."

EROSION (i) the wearing away of the land surface by rain or irrigation water, wind, ice, or other natural or anthropogenic agents that abrade, detach and remove geologic parent material or soil from one point on the earth's surface and deposit it elsewhere; (ii) the detachment and movement of soil or rock by water, wind, ice, or gravity.

HYDROSPHERE All the waters on the earth's surface, such as lakes and seas, and sometimes including water over the earth's surface, such as clouds (the water).

ILLUVIATION The translocation (deposition) of soil material in a soil horizon that has been removed from another horizon by eluviation; usually from an upper to a lower horizon in the soil profile.

LITHOSPHERE The solid, outermost surface of the earth (the rocks).

LEACHING The process of removal (loss) of dissolved soil materials (that are in solution in water filling soil pores), such as nutrients, out of the soil into the parent material and/or groundwater below it.

LOSSES The process of removal of soil materials by various pathways; i.e., leaching, erosion.

PEDOSPHERE That shell or layer of the earth in which soil-forming processes occur. It is where the lithosphere, biosphere, hydrosphere and atmosphere intersect (the soil).

SOIL A natural, three-dimensional body at the earth's surface. It is capable of supporting plants and has properties resulting from the effects of climate and living matter acting on earthy parent material, as conditioned by relief and by the passage of time.

TRANSFORMATIONS The process of chemical, physical or biological change of soil materials – when a component turns into a different component, such as through decomposition and/or weathering.

TRANSLOCATIONS The process of movement of soil materials within the soil—movement from one location in the soil to another (eluviation, illuviation).

CHAPTER 2

PHYSICAL PROPERTIES OF SOIL AND SOIL FORMATION

As we discussed in chapter 1, soil is composed of minerals, organic material, water, and air. Variation of these materials gives soil its physical, chemical, and biological properties. The next several chapters will describe these soil properties and discuss how they relate to each other and to the environment. This chapter begins by focusing on the physical aspects of soil, and ends with the process of soil formation.

One way to begin to understand soil is to simply dig a hole (figure 2–1). Looking at the exposed soil reveals several obvious physical aspects. Initially, you can see that soil has many colors. It is not simply brown or red or black. Next, you can see patterns to the color. The upper portion of the soil is darker (browner or blacker) than the materials below, and the colors appear in horizontal bands or layers. A closer examination reveals that plant roots are more common in the upper layers. Furthermore, the soil is composed of clumps, clods, or *aggregates* that are made up of still smaller particles. There is also a different feel from the top layer to the bottom in terms of how easy it is to dig or break apart. These simple observations identify many of the key physical characteristics, or morphology, of soil: color, texture, structure, consistence, and horizons (table 2–1). Soil scientists use these characteristics to create a soil's profile description (also referred to as the soil morphology or morphological description) (table 2–2).

SOIL COLOR

Color is perhaps the most obvious and easily determined soil property. You can infer from a soil's color extremely important site characteristics, such as drainage, mineral weathering, and water content (figure 2–2).

In surface layers, organic matter darkens the soil, usually masking all other coloring agents. In the subsoil, iron (Fe) is the primary coloring agent. The bright, uniform orange-brown colors often associat-

CHAPTER AUTHORS

DAVID LINDBO
WALE ADEWUNMI
RICH HAYES

ed with well-drained soils are the result of iron oxides that coat individual particles. Soils with a fluctuating water table usually have a mottled, or spotted, pattern of gray, yellow, and/or orange colors. Very poorly drained soils have a high water table for much of the year and have a very gray background. The term *gley* is used for such gray, poorly drained soils. Poorly drained soils pose significant land use challenges unless they are artificially drained.

Manganese (Mn) is common in some soils as manganese oxides and results in very dark black or purplish black spots or stains. Several other soil minerals have distinct colors, thus making their identification straightforward (table 2–3 for other typical colors and formulas). For example,

Figure 2–1. Various views of the soil as we look closer—(a) landscape, (b) the pit being described, c) the soil profile, (d) roots, (e) structure, and (f) coarse fragments and texture.

glauconite (common in green sands) is green, and quartz (common in granite) has various colors but is often white or gray. Feldspars (common in granite) range from pale buff to white. Micas (in granites and other igneous rocks) may be white, brownish black, or golden, and kaolinite (a secondary mineral) appears gray to white.

Granite is typically composed of quartz, feldspar, and mica.

Color determination can be quite subjective. In general, people perceive colors differently, so soil scientists use a standard reference known as the *Munsell color system*. (Munsell soil color books can be purchased from many research and industry supply catalogs; see chapter 5 for more on the history of the Munsell system.) The system describes three components of color: *hue*, the dominant wavelength of reflected light; *value*, the lightness or darkness of a color; and *chroma*, the relative strength of the hue (figure 2–3).

Determining the variation in color throughout the soil is important. The *matrix color* is the dominant color of the

Table 2–1. Two views of soil characteristics: general and scientific.

General observation	Soil science terminology
Pit or hole in the ground with lots of layers	Soil profile
Layers of color Topsoil Subsoil	Soil horizon
Difficulty to dig Sticky Gumbo clay	Consistence
Clumps Chunks Clods	Soil structure
Sandy Silty Gravelly Clayey	Soil texture
Black Topsoil Subsoil	Soil color—Munsell designation
Splotches Red mottles Tiger dirt Gray mottles	Mottles Redoximorphic features (concentrations, depletions)
Stones Gravel Rocks Boulders	Coarse fragments
Heavy soil Light soil	Bulk density
A hole in the ground	An opportunity to investigate the soil profile

Table 2–2. Example of a soil profile description.

	Slopes 10%. Aspect SW Landscape position: Side slope VL
A	0 to 25 cm; 5YR 4/3 sandy loam; >2% 5YR 4/2 mottles; single grained; loose
Bt1	25 to 70 cm; 2.5YR 7/4 sandy clay; 10% 7.5YR 5/8 redox concentrations; subangular blocky; friable; slightly sticky, slightly plastic
Bt2	70 to 135 cm; 2.5YR 7/6 clay; common medium distinct 7.5YR 6/2 redox depletions; subangular blocky; firm; moderately sticky; moderately plastic
BC	135 to 175 cm; 5YR 7/3 sandy clay; weak medium subangular blocky; firm; moderately sticky; moderately plastic
C	175 to 225cm; 7.5YR 8/2 loamy sand; massive-RCF; loose

Note: The terminology used in the description is defined throughout the chapter. The soil profile description is also referred to as the soil morphology.

group—these have low chroma matrix colors with or without redoximorphic features. As mentioned above, if the soil has a gray color, then it is likely to be wet for some portion of the year.

We can use the presence of gray, low chroma colors (≤2 chroma) in the soil to determine the water table depth. This is done for many land uses related both to agriculture and to urban development. These features also help soil scientists identify wetland or hydric soils. Many land use decision are based on these colors and the fact that they do not change from season to season. Thus in summer when water tables are deep, the gray colors indicate how high the water table will rise during the wettest time of the year, an important thing for builders and farmers to know.

SOIL TEXTURE

Pick up a handful of soil and you can feel how fine or coarse it is. That feel comes from the size and relative proportion of mineral particles in the soil, and is known as the soil *texture* (figure 2–5, tables 2–4 and 2–5). Many soil properties are related to the size of the mineral particles and their distribution. These properties include *bulk density, porosity, permeability, infiltration,* and *water-holding capacity,* which will be discussed later in this chapter, so it is important to understand how soil texture is determined and described.

Soil texture can be determined quantitatively in the laboratory or estimated in the field. The field method is also referred to as "texture by feel" (figure 2–6). This method works well if the practitioner has worked with known samples and practiced a great deal. A simpler approach is to break the soil into three general groups: sandy, loamy, or clayey. Sandy soils do not form a ball when a moist sample is squeezed in the hand. Loamy soils will form a ball when moist but will not form a ribbon of more than 5 centimeters when pushed between thumb and forefinger.

horizon, or layer (figure 2–4). Some horizons have several colors. Soil scientists record the dominant color first, followed in sequence by the minor colors. Mottles are zones or spots of color different from the matrix. These colors may result from the parent material (rock fragments), mineralogy, weathering patterns, concretions, nodules, cemented bodies, filled animal burrows, or root channels. A special group of mottles called *redoximorphic features* relate directly to soil wetness or saturation. Gley colors are another special

Soil Minerals and Soil Color

As we will learn later in this chapter, rocks are the parent material of soil, and rocks are composed of minerals. Minerals, in turn, are made up of elements such as silicon and oxygen, which combine to create quartz (SiO_2). Silicon and oxygen are the most abundant elements in Earth's crust, but other elements combine to form thousands of mineral types. The minerals present in rocks (primary minerals such as feldspar or biotite) weather in the soil to form new minerals (secondary minerals such as kaolinite, illite, and iron oxides). These minerals (both primary and secondary) are key ingredients in a soil's color. Along with black and dark brown organic matter, secondary minerals such as iron oxides and hydroxides are of most importance to the color of the soil.

Table 2–3. Properties of soil minerals.

Mineral	Formula	Color
Primary		
Quartz	SiO_2	light gray
Muscovite (mica)	$KAl_2(AlSi_3O_{10})(F,OH)_2$	silvery white or gray
Biotite (mica)	$K(Mg,Fe)_3AlSi_3O_{10}(F,OH)_2$	black
Feldspar (orthoclase)	$KAlSi_3O_8$	grayish yellow to white
Feldspar (albite)	$NaAlSi_3O_8$	white to light gray
Feldspar (anorthite)	$CaAl_2Si_2O_8$	white to gray
Primary and secondary		
Calcite	$CaCO_3$	white
Dolomite	$CaMg(CO_3)_2$	white
Gypsum	$CaSO_4 \times 2H_2O$	very pale brown
Secondary		
Goethite	$FeOOH$	yellow
Hematite	Fe_2O_3	red
Lepidocrocite	$FeOOH$	reddish-yellow
Ferrihydrite	$Fe(OH)_3$	dark red to yellow
Iron sulfide	FeS	black
Jarosite	$KFe_3(OH)_6(SO_4)_2$	pale yellow
Todorokite	MnO_4	black
Humus		black to dark brown

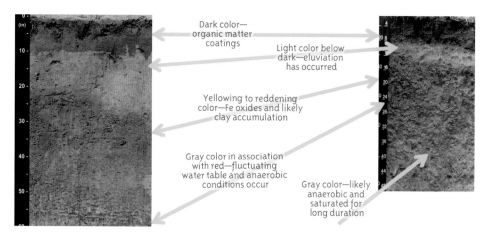

Figure 2–2. Color changes in a soil profile mean something. Dark color indicates organic matter coatings. Light color below dark indicate *eluviation* has occurred. Yellowing to reddening color indicate iron oxides and likely clay accumulation. Gray in association with red suggests fluctuating water table and anaerobic conditions. Gray color indicates likely anaerobic conditions, in which the soil has been saturated for a long duration.

Figure 2–3. Color is three-dimensional. Soil scientists use the *Munsell Soil Color Book* to identify soil color. *Hue* refers to the dominant wavelength of reflected light (e.g., red, yellow, green) and is designated as a number-letter sequence, i.e., 10YR or 10R. Value refers to the lightness or darkness of a color in relation to a neutral gray scale. Value extends from pure black (0/) to pure white (10/) and is a measure of the amount of light that reaches the eye. Gray is perceived as about halfway between black and white and has a value notation of 5/. Lighter colors have values between 5/and 10/; darker colors lie between 5/to 0/. Chroma is the relative purity or strength of the hue. Chroma indicates the degree of saturation of neutral gray by the spectral color. Chromas extend from /0 for neutral colors to /10 as the strongest expression of the color. The typical notation of color is an alpha-numeric term of hue value/chroma such as 10YR 5/6.

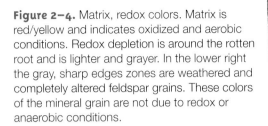

Figure 2-4. Matrix, redox colors. Matrix is red/yellow and indicates oxidized and aerobic conditions. Redox depletion is around the rotten root and is lighter and grayer. In the lower right the gray, sharp edges zones are weathered and completely altered feldspar grains. These colors of the mineral grain are not due to redox or anaerobic conditions.

Redoximorphic Features

Redoximorphic features result from chemical transformations (including reduction and oxidation, which are discussed in chapter 3) of iron and/or manganese. These color features are often used to identify the depth where saturation occurs in the soil. Two types of redoximorphic features commonly found in soils are redox depletions and redox concentrations.

Redox depletions are gray (low chroma) areas in the soil that formed due to prolonged saturation. These features are used by soil scientists to identify the presence and depth of seasonal high water tables in a soil. When soil becomes saturated, soil microbes utilize all the dissolved oxygen, resulting in anaerobic conditions. Under anaerobic conditions the iron in the iron oxide is biochemically transformed (reduced) through a microbial reaction. The reduced iron is then translocated as the water it is dissolved in moves through the soil. With the thin red, brown, yellow, or orange colored iron coatings removed, the soil color changes to that of the underlying mineral grains, which is often gray or white in color. The reaction only occurs in saturated conditions (i.e., where all soil pores filled with water) and if there are sufficient numbers of microbes and a food source (carbon) that allows them to use up the dissolved oxygen in the soil water to create anaerobic conditions (no oxygen present in the soil water).

Redox concentrations are areas of iron and manganese oxide accumulation that formed as a result of chemical reduction and oxidation reactions. If reduced iron moves to an area where oxygen is dissolved in the water or next to a pore that has air in it, the reduced iron will oxidize, or rust. This results in the area of the redox concentration containing more iron than the surrounding soil. The increase in iron will change the color of the soil, resulting in a higher chroma color. Redox concentrations are commonly in shades of red, brown, yellow, or orange.

TYPICAL OXIDATION AND REDUCTION REACTIONS

If air (O_2) is in the soil the soil is aerobic
$4e^- + O_2 + 4H^+ \rightarrow 2H_2O$;
rusty or oxidized color persist

If all O_2 is removed soil becomes anaerobic (saturation occurs)

Denitrification
$10e^- + 12H^+ + 2NO_3 \rightarrow N_2 + 6H_2O$;
no color change

Iron reduction
$2e^- + 6H^+ + Fe_2O_3 \rightarrow 2Fe(II) + 3H_2O$;
soil turns gray

Sulfate reduction
$8e^- + 10H^+ + SO_4 \rightarrow H_2S + 4H_2O$;
rotten egg odor

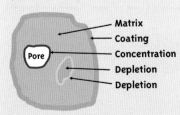
Matrix — Coating — Concentration — Depletion — Depletion (Pore)

1. Plant root grows into soil

2. Root dies and starts to decompose

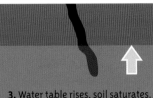
3. Water table rises, soil saturates. Bacteria use up oxygen in water around the root.

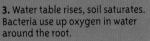
4. Soil is saturated. Bacteria continue to decompose root and reduce iron around root. This turns the soil near the root gray.

5. Soil is still saturated. Reduced iron diffuses (moves) away from root, oxidizes and the soil turns red.

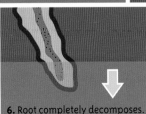
6. Root completely decomposes. Water table drops.

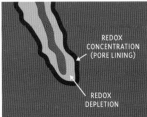
REDOX CONCENTRATION (PORE LINING)
REDOX DEPLETION

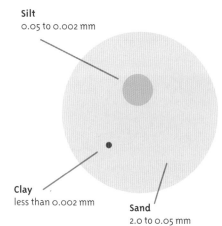

Figure 2–5. Relative sizes of gravel, sand, silt, and clay particles.

Table 2–5. Soil particle size and resultant surface area.

Particle	Diameter	Volume	Number of particles per gram	Surface area in 1 gram
	mm	mm³		cm²
Very coarse sand	2.00–1.00	4.18	90	11
Coarse sand	1.00–0.50	0.524	720	23
Medium sand	0.50–0.25	0.0655	5700	45
Fine sand	0.25–0.10	0.00818	46,000	91
Very fine sand	0.10–0.05	0.000524	722,000	227
Silt	0.05–0.002	0.0000650	5,776,000	454
Clay	<0.002	0.0000000042	90,260,853,000	8,000,000

Table 2–4. Absolute sizes of gravel, sand, silt, and clay.

Name	Size	Feel
	mm	
Gravel	>2	very gritty to pebbly
Sand	2–0.05	gritty—sand paper
Silt	0.05–0.002	floury, soft, smooth
Clay	<0.002	sticky, slick

The sizes shown can be further subdivided into the fine-earth fraction, those <2 mm (sand, silt, clay) and the coarse fragments, or rock fragments, which are all material >2 mm (gravel, stones).

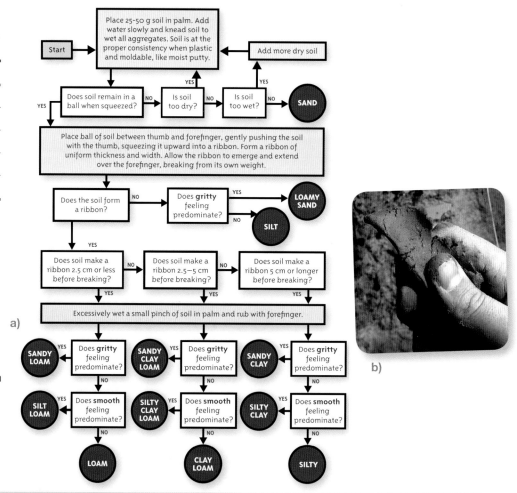

Figure 2–6. Soil scientists learn to determine soil texture by feel. (a) Use the flow chart to determine the texture. (b) Can you make a "ribbon" of soil?

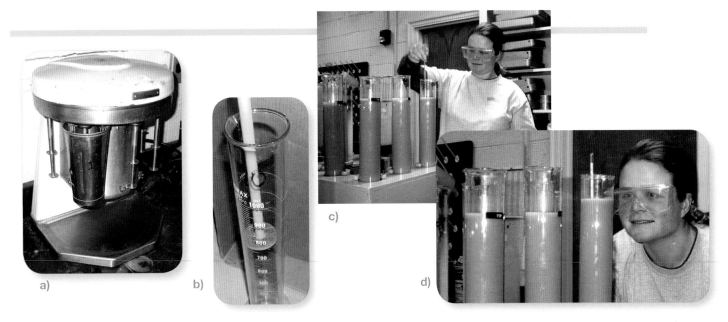

Figure 2–7. Particle size in soil is determined most precisely in the lab. (a) A known weight of soil is mixed (yes it is a milk shake mixer—but for soil only!) with a chemical (called a "deflocculant") to make sure the particles do not clump ("flocculate"). (b) The mixture is placed in a settling cylinder, the volume is brought to 1000 milliliters, and at the start of the procedure the soil and water mixture is mixed again with a plunger. The plunger is immediately removed, and particles are allow to settle. (c) The density, and thus the amount of specific particles in the water, is measured using an instrument called a hydrometer. (d) The sand can be sieved out of the liquid once the clay measurement has been taken. The sand is then sieved to separate it into very coarse to very fine sand.

Clayey soils also form a ball when moist and will also form a ribbon of greater than 5 cm when forced between thumb and forefinger. The laboratory method is more quantitative and uses the relative rates at which different sized particles fall or settle in a liquid due to gravity (figure 2–7).

Whether texture is determined by feel or in the laboratory, the U.S. Department of Agriculture and soil scientists in general recognize 12 soil textures (aka USDA textures) based on the percentages of sand, silt, and clay (table 2–6). The texture triangle (figure 2–8) is used to determine the name of the soil texture. When referring to loam one is only indicating that the soil has a given percentage of sand, silt, and clay; there is no mention of whether or not that soil has any organic matter in it or not.

Beginning with the largest mineral particles, coarse fragments are greater than 2 millimeters in diameter, and are primarily composed of unweathered or slightly weathered rock fragments such as gravel, stones, and cobbles. They are relatively inert in the soil and are not considered directly when determining the USDA soil texture (figure 2–8, table 2–5). If coarse fragments are present in amounts exceeding 15%, an adjective is added to the texture. For example, if a sandy loam soil contains 15%–35% gravel it will be called a gravelly sandy loam, 35%–60% very gravelly, and >60% extremely gravelly. The presence of the coarse fragments can affect soil properties by lowering porosity, permeability, and water-holding capacity while increasing bulk density and making the soil more difficult to till or work for agricultural and nonagricultural uses.

Next in size are sand and silt particles. Like coarse fragments, they may be unweathered or slightly weathered rock fragments containing primary minerals. They may also be made up of resistant primary minerals such as quartz that remain from the breakdown of larger rocks. Together sand and silt form the skeleton of the soil.

Table 2–6. Textures and ranges of sand, silt, and clay.

Texture	Sand	Silt	Clay
		%	
Sand	80–100	0–15	0–10
Loamy sand	70–85	0–30	0–15
Sandy loam	50–85	0–50	0–20
Loam	25–52	28–50	7–26
Silt loam	0–50	50–90	0–26
Silt	0–20	80–100	0–10
Sand clay loam	45–80	0–28	20–35
Clay loam	20–45	16–54	26–40
Silty clay loam	0–26	40–74	26–40
Sandy clay	45–65	0–20	35–55
Silty clay	0–20	40–60	40–60
Clay	0–45	0–40	40–100

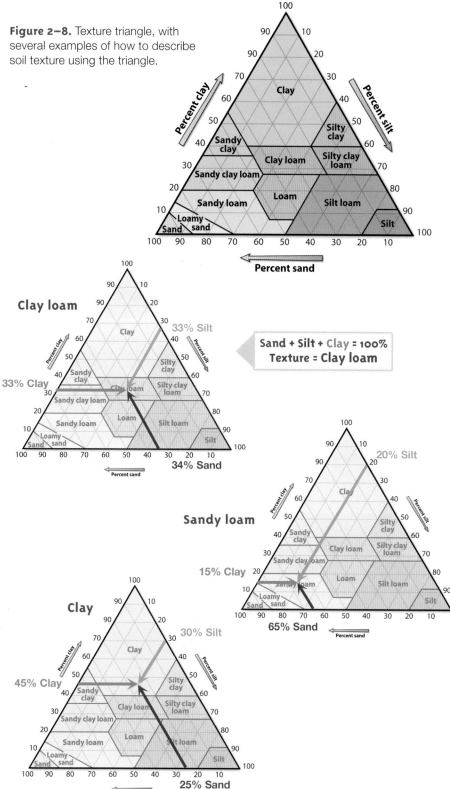

Figure 2-8. Texture triangle, with several examples of how to describe soil texture using the triangle.

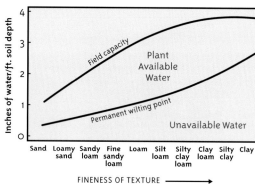

Figure 2-9. A general graphical representation of the amount of water available for plant use (plant available water) is shown as the area between the field capacity line (water held in the soil after gravitational water has been removed) and the permanent wilting point line (water that cannot be extract by plant roots). The largest amount of plant available water occurs when the soil texture is in the loam to silty clay loam range. Adding organic matter will increase the amount of plant available water.

Soils with high sand contents tend to have large pores and therefore allow gases and water to move through them rapidly. Silt particles are relatively inert (as are sand particles) and have smaller pore spaces, resulting in slower movement of gasses and water. The smaller pore space in silts results in more water being held in these soils, therefore more water is available for plant growth (figure 2–9).

Finally, clay particles are so small they are not visible to the naked eye yet their properties control many other soil properties. Clay particles are characterized not only by size but also by their shape and crystal structure. Unlike sand and silt particles, which are often slightly weathered primary minerals, clays are most often secondary minerals, which are minerals that have formed in soil as opposed to being inherited from the parent material or bedrock.

Clay particles have a very large surface area compared with their volume (table 2–7). As the particle diameter decreases, the surface area increases dramatically. For example, a pea-sized volume weighs approximately 1 gram (0.035 oz.). If this amount was sand its surface area would be about one-fifth to one-quarter the size of a credit card. The same volume of silt would have approximately the equivalent surface area as one side of a sheet of paper. Lastly, a pea-sized volume of clay would have a surface area as large as a football field. This large surface area is due to the platy nature of clay, resulting in large internal surface area of the clay particles.

The surface area of a particle is related to its chemical reactivity; sand is much less reactive than clay (table 2–8). The importance of this reactivity to soil's physical and chemical properties is profound and is discussed here as well as in subsequent chapters.

Clay's negative charge and reactivity give it the plasticity and cohesiveness that allow it to be molded into pots and bowls. Because of their stickiness when wet, soils high in clay are worked only with difficulty. Water and air move many times more slow-

Table 2–7. Clay mineral properties.

Clay mineral	Type	Interlayer condition/ Bonding	Cation exchange capacity	Swelling potential	Specific surface area
			cmol/kg		m²/g
Kaolinite	1:1 (non-expanding)	lack of interlayer surface, strong bonding	3–15	almost none	5–20
Illite (hydrous mica)	2:1 (non-expanding)	partial loss of K, strong bonding	10–40	low	50–200
Montmorillinite (Smectite)	2:1 (expanding)	very weak bonding, great expansion	80–150	high	700–800

The Chemistry of Clay

Chemically, the reactivity of clay is due not only to the large surface area but also to its negative charge. This charge occurs for several reasons. One is the nature of the crystal structure of the clay (how the particles are built on an elemental level). Clay has an *isomorphic substitution* of aluminum for silicon in the tetrahedral positions in the mineral structure. It also has broken bonds that occur between the clay layers or at the edges of the crystal structure where the mineral has been physically disturbed or damaged. Finally, there is a pH dependent charge associated with hydroxyl groups (–OH) that will lose the hydrogen ion (H+). The negative charge attracts positively charged atoms (cations) of nutrients such as calcium, potassium, magnesium, zinc, and iron.

Types of Clays

There are many types of clay minerals, which are plate-shaped aluminosilicates (phylosilicates) Of all the types of clays kaolinite, smectite, and illite are the most important. The smectite clays are often referred to as expansive, or shrink–swell clays, meaning they swell when wet and shrink when dry in the same way a sponge does. Smectite clays are much more reactive (physically and chemically) than the non-expansive kaolinite or illite clays. Kaolinite is most often observed in soils that are highly weathered, whereas illite occurs in slightly weathered soils. However, all three major clay minerals may be found in a given soil. Most pottery clay is kaolinitic or illitic since it does not shrink as much when it dries. Smectite clays are often used to line ponds to retain the water since when they swell they are nearly impermeable to water.

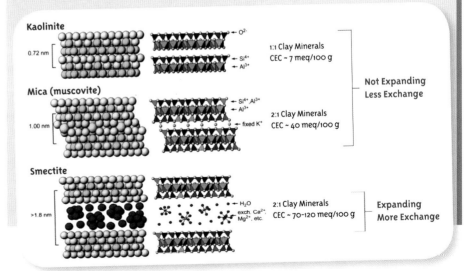

Table 2–8. Relations of sand, silt, and clay to bulk density, porosity, pore size, permeability, plant available water, erodibility etc.

Property/behavior	Sand	Silt	Clay
Water holding capacity	low	medium to high	high
Plant available water	low	high	medium
Aeration	good	medium	poor
Organic matter decomposition	fast	medium	slow
Organic matter content	low	medium to high	high to medium
Water erosion	low	high	low
Sealing for ponds, landfills	poor	poor	good
Nutrient supplying	poor	medium to high	high
Pollutant leaching	high	medium	low
Drainage rate	high	medium to slow	slow to very slow
Wind erosion	moderate to high if very fine sand	high	low if well structured, high if not
Shrink–swell	low	low	moderate to high depending on clay minerals
Porosity	low	medium	high
Permeability	high	medium to low	low to very low
Bulk density	high	medium	low
Potential for smearing if tilled when wet	low	medium	high
Compactability	low	medium	high
Warming in spring	fast	moderate	low
Infiltration	fast	medium	low

ly through clay than through sand, even though the total amount of pore space in clay is greater than the pore space in sand. This is because the diameter of the pores in clay is very small. In fact, a portion of the water held by clay is bound so tightly that most root systems cannot absorb it, which makes soil composed of very high amounts of clay poorly suited for growing crops.

SOIL STRUCTURE

Another physical characteristic of soil is its structure. How does structure differ from texture? Soil *structure* is the arrangement of individual soil particles (texture) into a larger grouping, sometimes called *peds* or *aggregates*. One can also think of structure as a brick house, in which the brick, mortar, and cement are the particles (i.e., texture) and the completed geometry of the house is the structure. Consider a soil with a type of structure that does not allow roots to penetrate deeply into the soil—what will be the effect on crops, trees, and runoff? Agricultural crops will suffer water stress because the plants will be unable to use water deep in the soil profile. In a forest, shallow rooting because of poor structure can result in trees that are easily blown over during wind gusts. If rainwater cannot penetrate the soil surface because of poor structure, runoff will occur. If this is excessive and other conditions are right, severe erosion can follow.

What makes for good or poor soil structure? A well-structured soil has an abundance of pores for water and air to move readily into the soil and to root systems. The result is that after a rainfall, infiltration is enhanced so that water penetrates the soil instead of running off and carrying away sediments. The aggregates of soil with good structure are more difficult to move and so are resistant to the erosive effects of wind and water. Overall, good structure enhances crop production, soil aeration, root growth, infiltration, and stability. Unlike soil texture, soil structure can be altered. Soil structure can be improved by managing fields to include cover crops and no-till planting.

A poorly structured soil has little pore space for air and water because the soil aggregates are packed tightly or they may not even exist. Individual soil aggregates are unstable and fall apart readily when disturbed. Water infiltration may be greatly reduced, and rainwater or irrigation water may not penetrate deeply enough to sustain crops. The low amount of pore space can cause ponding in low-lying sites or result in runoff and subsequent erosion on upland sites.

Soil structure forms naturally through many processes, including alternate freezing and thawing; wetting and drying; plant root penetration; burrowing by worms, insects, and other animals; the addition of secretions or exudates from plants and animals; and the microbial decay of plant and animal remains. Adding organic matter, lime, and manure often increases soil productivity by improving its aggregation or structure.

Soil structure can be degraded by compaction from equipment and even foot traffic (figure 2–10). Even the impact of raindrops on unprotected soil can degrade soil structure. The susceptibility of a soil to compaction changes with its moisture content. Dry soils resist compaction while wet soils are vulnerable to compaction. For this reason, it is wise to schedule any activities that may compact the soil to times of low soil moisture. Soil structure in the A

Figure 2–10. Foot traffic compaction destroys soil structure and is killing these trees, causing or enhancing erosion.

Types of Soil Structure

Soil structure is defined as the naturally occurring arrangement or grouping of a soil's primary particles (sand, silt, and clay) into aggregates. These aggregates are classified according to their shape: granular, angular blocky, subangular blocky, platy, wedge, prismatic, and columnar. Soils that do not have aggregates or structure are referred to as *structureless*.

Typically, granular structure looks like granola and is most often found in surface layers with appreciable organic matter levels. Angular blocky structure is equidimensional with the faces at sharp angles and the peds fitting together well. Angular blocks look like new fresh building blocks before their edges have become rounded due to use (then they are subangular blocky).

blocky

Subangular blocky structure has more rounded than angular corners and edges. Platy structure is characterized by horizontal planes that look like plates. Wedge structure has elliptical interlocking peds thate often show faces that appear smooth. Wedge structure looks like a block that has been pulled at opposite ends, resulting in an elongated block. Prismatic structure is vertically elongated. Columnar structure is similar to prismatic, but the unit tops are frequently rounded and bleached. Columnar structure is associated with horizons that are high in salts, which cause this type of structure to form.

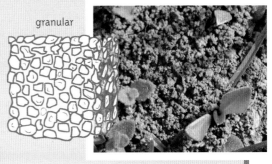
granular

There are three types of structureless soil: *massive, massive-rock controlled fabric*, and *single-grained*. Single-grained means there is no cohesion between soil particles, such as sand at the beach. Single-grained refers to non-cohesive sands, whereas massive refers to any soil that does not break into any predictable and repeatable type or shape.

Massive means that there is no arrangement of soil particles into "real" structural units. This is often found deep in the soil or when the soil particles are cemented together. Another type of massive structure is called massive-rock controlled fabric. Massive-rock controlled structure describes soil developed from saprolite, which is weathered bedrock typically found in the Piedmont of the southern United States. Unlike simple massive structure, massive-rock controlled fabric may have a preferred orientation due to the minerals present in the parent rock. The material may easily break into the individual mineral grains and may not be as limiting to water movement or root growth as a truly massive soil structure.

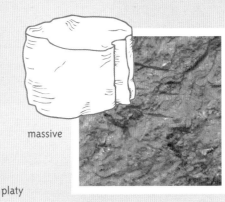
massive

single-grained

columnar or prismatic

platy

horizon of a undisturbed land and cultivated land are quite different just because of the disturbance by plowing year-in-year-out. However, plowing the topsoil for planting loosens it and increases water infiltration.

SOIL CONSISTENCE

In describing soil texture we often refer to clay as being sticky, moldable, or plastic. The terms sticky and plastic are also used in describing another soil property: soil *consistence*. Soil consistence is somewhat related to texture as soils with the greatest consistencies are often high in clay content. Consistence is a description of how well soil holds together (rupture resistance or moist consistence), the ease with which it deforms (plasticity), and how sticky it is (stickiness). We'll discuss these three ways to describe soil consistence in this section (table 2–9).

Rupture resistance is a measure of the strength of the soil material to withstand an applied stress. It can be determined under either dry or moist conditions but is most often measured moist. The more stress that is needed to deform the ped, the stronger the soil or the greater its consistence or rupture resistance is. A soil with a high rupture resistance will be more difficult to till or excavate. A soil with a high rupture resistance may have high expansive clay content or be cemented.

Soil stickiness is the capacity of soil material to adhere to other objects. Stickiness is estimated at the moisture content that displays maximum adherence between thumb and forefinger (figure 2–11). As the amount of expansive clay in the soil increases, the stickiness is also likely to increase. A soil with a high stickiness will be difficult to till or work when wet. Very sticky soils can become impassable to vehicles and pond water. They also have a low hydraulic conductivity.

Soil plasticity is the degree to which soil material can be permanently deformed without rupturing. The evaluation is made by forming a roll of soil at the water content where the maximum plasticity is ex-

Table 2–9. Soil consistence is broken into moist consistence (rupture resistance) and wet consistence (stickiness and plasticity). Moist consistence is performed on a natural soil ped or aggregate, wet consistence is done on disturbed or mixed sample where water has been added. The exact amount of water varies from soil to soil and is related to the types and amounts of clay minerals present in the sample.

Name	Description of how a ped or aggregate responds to pressure when squeezed
Rupture resistance or moist consistence	
Loose	cannot get sample
Very friable	very slight finger force (squeezed between thumb and forefinger)
Friable	slight finger force—like really rich, crumbly cake
Firm	moderate finger force—like toast
Very firm	strong finger force—like very stale bread
Extremely firm	moderate hand force (squeezed between hands)
Slightly rigid	foot pressure (step on the sample)
Rigid	light blows (hit the sample with a hammer)
Very rigid	strong blows
Stickiness	
NAME	DESCRIPTION OF HOW A DISTURBED SAMPLE STICKS TO FINGERS
Non-sticky	little or no soil adheres to fingers after release of pressure
Slightly sticky	soil adheres to both fingers after release of pressure with little stretching on separation of fingers
Moderately sticky	soil adheres to both fingers after release of pressure with some stretching on separation of fingers
Very sticky	soil adheres firmly to both fingers after release of pressure with stretches greatly on separation of fingers
Plasticity	
NAME	DESCRIPTION OF HOW A DISTURBED SAMPLE FORMS A "WIRE" WHEN ROLLED OUT
Non-plastic	will not form a 6-mm-diameter, 4-cm-long wire, or if formed, cannot support itself if held on end
Slightly plastic	6-mm-diameter, 4-cm-long wire supports itself, 4-mm-diameter, 4-cm-long wire does not
Moderately plastic	4-mm-diameter, 4-cm-long wire supports itself, 2-mm-diameter, 4-cm-long wire does not
Very plastic	2-mm-diameter, 4-cm-long wire supports itself

pressed (figure 2–11). A highly plastic soil can be molded into different shapes and will retain those shapes when dry. As with a very sticky soil, very plastic soils often have high expansive clay content and are difficult to work when wet and often have a low hydraulic conductivity.

Consistence also relates to other soil properties and is most critical to water movement and overall soil management. Aspects of consistence determine if the soil has low permeability. Essentially, as consistence increases, the rate of water movement decreases.

Figure 2–11. Sticky vs. non-sticky soil. (left) Non-sticky—the soil adheres to only one finger. (right) Very sticky—it's everywhere!

SOIL HORIZONS AND PROFILE

So far we have focused on the characteristics of a handful of soil. Soils, however, do not exist as small discrete handfuls but as complex natural materials across the landscape. To communicate to other soil scientists and professionals, soil scientists have developed a uniform method to describe soils. This begins with digging a hole and observing the differing colors and layers, or horizons.

Soil *horizons* are approximately parallel to the surface and have formed in place as a result of the soil processes introduced in chapter 1. You can easily see soil horizons in excavations or road cuts. A well-developed or older soil may have many horizons whereas a young or poorly developed soil may have only two visible horizons. In areas of significant soil or land disturbance, such as urban areas, where erosion is severe, or where landslides have occurred, some horizons will be missing. A soil profile is a vertical cross-section of all the soil horizons at a particular location and forms the basis for understanding and communicating all the soil properties.

Soil horizons are described using a series of letters and numbers. In this book we focus on the most common horizons (table 2–10). Capital letters designate *master horizons*: O, A, E, B, C, and R horizons. The thickness of each layer varies with location. Lowercase letters are used as suffixes to indicate specific characteristics of the master horizon and indicate soil processes associated with these characteristics. These are the *subordinate horizon* designations. For example, an Ap horizon is an A horizon that has been plowed (p). Arabic numerals are used as suffixes to indicate vertical subdivisions within a horizon. For example, if there are changes in color from the top of the E horizon to its base, the designation would be E1 and E2 based on that change in color.

The O horizon (organic horizon), though not always present, is generally the uppermost layer of the soil and is made up of *organic material*. It consists of accumulations of organic matter in various stages of decay. This horizon is often found in undisturbed areas such as wetlands and forests. It is not present in agricultural fields or in suburban or urban settings where the soil has been disturbed by plowing or development.

The A horizon is commonly called *topsoil* and typically has friable consistence (table 2–8) and a granular structure. It is usually darker than the lower layers because of its higher organic matter content, and is often more fertile when compared with underlying horizons. Also known as the plow layer in cultivated fields, this horizon is where most root activity occurs.

When the plow layer is obvious in the profile with a smooth abrupt boundary, it is designated as an Ap horizon. Ap horizons may remain obvious for decades after annual plowing has ceased. This horizon is subject to materials being dissolved and translocated deeper in the profile.

The E horizon, or eluvial horizon, is characterized by a light color or bleached appearance and is a zone of removal, or eluviation (materials exiting). Dissolved minerals, nutrients, and clay move out of the horizon carried by water as it percolates downward in the profile. The primary feature of this horizon is loss of clay, iron, aluminum, and/or organic matter leaving a concentration of more inert sand and silt particles. It is commonly found in well-developed soils or soils that have undergone extreme translocation or leaching. It is less common in plowed areas where the E horizon tends to mix with the Ap horizon.

The B horizon is also referred to as the illuvial horizon (zone of accumulation), or subsoil. This horizon is usually higher value (lighter) than the A horizon due to its lower organic matter content. It is the zone of accumulation for materials eluviated from the A and E horizons. B horizons always include a subordinate horizon distinction. In well-developed soils the B horizon commonly has the highest clay content and is designated a Bt horizon. If the horizon simply involves a change in color or structure from that above or below it is designated a Bw horizon. When the B horizon has formed under conditions of long-term saturation and is gray in color it is a Bg horizon. (See table 2–10 for more subordinate designations.)

The C horizon is referred to as the parent material or the soil material similar to the parent material. It is less weathered than the upper horizons and does not have soil structure. It may be massive, massive-rock controlled fabric, or single grained. Often it contains partially disintegrated or weathered parent material transported by gravity, wind, and water or from the underlying bedrock. C horizons do not always have a subordinate horizon distinc-

Table 2–10. Horizons and description of each, along with a few of the subordinate horizons.

Master horizons and layers

	O horizons	Layers dominated by organic material.
	A horizons	Mineral horizons that formed at the surface or below an O horizon that exhibit obliteration of all or much of the original rock structure and (i) are characterized by an accumulation of humified organic matter intimately mixed with the mineral fraction and not dominated by properties characteristic of E or B horizons; or (ii) have properties resulting from cultivation, pasturing, or similar kinds of disturbance.
	E horizons	Mineral horizons in which the main feature is loss of silicate clay, iron, aluminum, or some combination of these, leaving a concentration of sand and silt particles of quartz or other resistant materials.

	B horizons	Horizons that formed below an A, E, or O horizon and are dominated by obliteration of all or much of the original rock structure and show one or more of the following:
		  illuvial concentration of silicate clay, iron, aluminum, humus, carbonates, gypsum, or silica, alone or in combination;
		  evidence of removal of carbonates;
		  residual concentration of sesquioxides;
		  coatings of sesquioxides that make the horizon conspicuously lower in value, higher in chroma, or redder in hue than overlying and underlying horizons without apparent illuviation of iron;
		  alteration that forms silicate clay or liberates oxides or both and that forms granular, blocky, or prismatic structure if volume changes accompany changes in moisture content; or
		  brittleness;
		  formation of pedogenic structure

	C horizons	Horizons, excluding hard bedrock, that are little affected by pedogenic processes and lack properties of O, A, E, or B horizons. The material of C horizons may be either like or unlike that from which the solum presumably formed. The C horizon may have been modified even if there is no evidence of pedogenesis. A Cr horizon is weathered, soft bedrock that can be dug by hand. A Cr horizon must not slake in water. If it slakes in water it should be considered a C horizon.
	R layers	Hard bedrock including granite, basalt, quartzite, and indurated limestone or sandstone that is sufficiently coherent to make hand digging impractical.

Subordinate distinctions within master horizons and layers (partial list)

d	Physical root restriction, either natural or manmade such as dense basal till, plow pans, and mechanically compacted zones.
g	Strong gleying in which iron has been reduced and removed during soil formation or in which iron has been preserved in a reduced state because of saturation with stagnant water.
h	Illuvial accumulation of organic matter in the form of amorphous, dispersible organic matter-sesquioxide complexes.
k	Accumulation of pedogenic carbonates, commonly calcium carbonate.
n	Accumulation of sodium.
o	Highly oxidized horizon composed of Al and Fe oxides with low silica contents.
p	Plowing or other disturbance of the surface layer by cultivation, pasturing or similar uses.
r	Weathered or soft bedrock including saprolite; partly consolidated soft sandstone, siltstone or shale; or dense till that roots penetrate only along joint planes and are sufficiently incoherent to permit hand digging with a spade.
s	Illuvial accumulation of sesquioxides and organic matter in the form of illuvial, amorphous, dispersible organic matter-sesquioxide complexes if both organic matter and sesquioxide components are significant and the value and chroma of the horizon are >3.
t	Accumulation of silicate clay that either has formed in the horizon and is subsequently translocated or has been moved into it by illuviation.
w	Development of color or structure in a horizon but with little or no apparent illuvial accumulation of materials.
x	Fragic or fragipan characteristics that result in genetically developed firmness, brittleness, or high bulk density.
y	Accumulation of gypsum.
z	Accumulation of salts more soluble than gypsum.

tion. The most common are the Cr designation, for a C horizon composed of slightly weathered bedrock; the Cd designation, for a dense C horizon (glacial till); or the Cg designation, if the horizon is gray and saturated for long duration.

The R horizon is bedrock. Unlike the Cr horizon, a backhoe usually cannot dig through it. Bedrock can be within a few inches of the surface or many feet below. It is often too deep to observe in a soil pit.

Occasionally, a horizon has properties of two adjacent master horizons. Soil scientists note this in the profile description through the use of specific transitional horizons. If the horizon is dominated by properties of one master horizon but with subordinate (lesser) properties of another, two capital letter symbols are used in sequence, such as AB, EB, BE, or BC. The master horizon symbol is given first and designates the horizon properties that dominate the transitional horizon. For example, a subsurface horizon below the A horizon may exhibit the color and texture of the A horizon above it, while having structure similar to the B horizon below. In this case, the horizon would be referred to as an AB horizon.

On the other hand, if the transitional horizon has discrete recognizable properties of the two kinds of master horizons then two capital letters are separated by a slash (/): E/B, B/E, or B/C. The first symbol is for the horizon that makes up the greater volume. Further changes to horizon names can be made after all the field properties are described, so the designation accurately describes the horizon.

So far we have focused on the more straightforward physical parameters of the soil (including horizons, color, texture, structure, and consistence). The next sections delve into the interactions between the physical makeup of the soil (minerals, pore space, organic matter) and water movement and related physical properties.

BULK DENSITY

Bulk density is defined as the weight of a material (in our case soil) divided by its volume. It stands to reason that something with high bulk density is "heavier" than something with low bulk density. People often use the descriptions "heavy" or "light" when talking about soil, and clay is typically described as a heavy soil, but this is a misconception (table 2–11). The misconception arises from the water-holding capacity of clay. Since clay holds more water and is inherently stickier than sand, it would take farmers more horsepower to plow a clayey field than a sandy field. This generated the common notion that clay soil was "heavier." Clay soils actually have a lower bulk density and higher porosity than sandier soils (table 2–11). This is due to the greater surface area of the clay particles, as discussed earlier. Although sand has large pores, there are fewer of them as compared with clay.

WATER AND WATER MOVEMENT

As mentioned, soil properties such as texture, structure, and the nature of the pores (and the bulk density) have a great influence on water movement through a soil. From a soil's perspective, the water cycle begins with rain falling on the soil surface where it can soak in (infiltrate), run off the soil surface, or evaporate (figure

Table 2–11. Bulk density, porosity, and texture relations. Bulk density is inversely related to the porosity. A high bulk density means there are more solid particles than pore space; sand with a high bulk density has a lower porosity than clay. This shows that clay is not in fact a "heavy" soil as its bulk density is lower than that of sand. However, even though clay has a high porosity the pores are small when compared to sand. The result is water movement in clay is slow due to the small pores and rapid in sand due to the large but fewer pores.

Texture	Bulk density	Porosity
	g/mL	%
Sand	1.55+	42
Sandy loam	1.40+	48
Loam	1.20	55
Silt loam	1.15	56
Clay loam	1.10	59
Clay	1.05	60

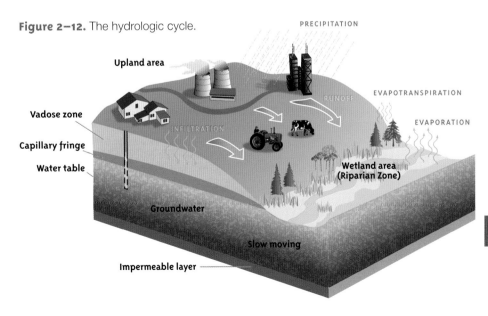

Figure 2–12. The hydrologic cycle.

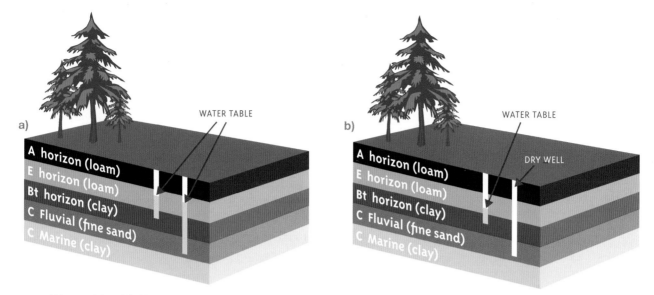

Figure 2–13. Water tables. (a) All wells are filled from bottom up. This is called "endosatuatrion." (b) Saturation in shallow wells but not in the deep well because the water is above the clay horizon—this is called "episatuartion" or a "perched water table."

2–12). Water that infiltrates moves downward through the soil pore spaces. In the root zone, water that is absorbed by roots moves through the plant to the leaves, where it is given off to the atmosphere as water vapor via *evapotranspiration*. In the atmosphere, water vapor condenses into clouds, then forms water droplets that fall back to the soil surface as rain, sleet, snow, hail, and dew. Water not used by plants continues to move through the soil flowing laterally—eventually surfacing as a spring or seep—or downward, eventually recharging the *groundwater*.

Precipitation that does not infiltrate moves across the surface as runoff, eventually entering streams, rivers, and ultimately the ocean. Whenever we open a tap we alter this pattern temporarily, but in the long run the water will rejoin the cycle. It is important to remember that the amount of water on Earth is constant, and soil acts as the primary filter to remove pollutants that are added to the water. Understanding the movement of water through soils is important for determining a soil's suitability for all types of land uses, whether it is agriculture, forestry, or urban development.

Water in the soil falls into two categories: *gravitational water* and *capillary water*. Gravitational water is free water that flows through (infiltration) or off the soil (runoff) as a result of the force of gravity. Gravitational water is responsible for moving nutrients (eluviation and leaching) and eroding soil. The loss of nutrients (leaching) due to gravitational water can be minimized by applying nutrients (fertilizer) in the correct amounts at the time when plants need them. If done incorrectly nutrients could be leached below the plant root zone. This is most important with sandy soils, as they are more susceptible to leaching losses. Irrigated areas are also susceptible to leaching and erosion.

Capillary water, unlike gravitational water, does not flow in the soil. It is present around individual soil particles and moves due to *capillarity*. This is a combination of the attractive forces between a water molecule and a surface (adhesion) and the attractive forces between water molecules (cohesion). Plant roots can use capillary water until the attraction between the water molecules and the soil particles becomes stronger than the pull of the roots. When water is not available the plant reaches the permanent wilting point. If water movement and extraction are slow, the plant may reach a temporary wilting point. This is why plants may appear to droop on hot days and then spring back once the temperature drops in the evening.

Water in the soil can also be described based on the pattern of saturation. *Saturation* is defined as the condition when all the soil pores are filled with water. Soils contain two distinct types of saturated zones, or *water tables* (figure 2–13): the apparent water table (*endosaturation*) or perched water table (*episaturation*). Endosaturation is when saturation is continuous from the top of the water table down. Episaturation is when a zone of saturation sits above a zone where the soil is unsaturated. This happens when a restrictive or impermeable horizon in the soil prevents water from moving deeper in the soil, and causes water to accumulate and saturate the soil above the impermeable horizon.

Each type of water table is associated with a *capillary zone* (or *capillary fringe*). The capillary zone occurs when water from a water table moves upward through the soil against the force of gravity because of the capillary attraction of water to the soil particles. Close to the true water table almost all the pores in the capillary zone are filled with water. As one moves further up from the water table the amount of water in the capillary zone decreases. The capillary zone is thickest in clays and thinnest in sands because of the pore size.

DEPTH TO SEASONAL SATURATION

As noted earlier, the water table is the zone in the soil that is saturated with water for significant periods during the year. However, the water table is not static; it moves up and down throughout the year depending on precipitation, runoff, interflow, and plant growth, among other factors. The seasonal high water table represents the highest point in the soil where the water rises during the wettest period of the year. The seasonal high water table is estimated by the presence of redoximorphic features (most commonly the first occurrence of ≤2 chroma depletions). Soil scientists have developed a rating table to describe where the seasonal high water table occurs in a given year (table 2–12). They also describe the frequency and duration of saturation in the soil according to *agricultural drainage class* (table 2–12).

Both of these measures—the depth to seasonal high water table (free water) and agricultural drainage classes—are used to identify the characteristics of saturation in the soil, or the *internal drainage* of the soil. Although confusing, agricultural drainage

Table 2–12. The depth to free water or water tables and agricultural classes. Saturation is determined by redoximorphic features or actual measurements. For agricultural drainage classes the exact range will vary state to state. Numbers presented here are an example.

Depth to free water or water table	Depth
Very deep	>150 cm
Deep	100–150 cm
Moderately deep	50–100 cm
Shallow	25–50 cm
Very shallow	0–25 cm

Agricultural Drainage Class	Depth range (example)	Description
Excessively drained (ED)	>120 cm and most often highly permeable sand	Water is removed very rapidly. The occurrence of internal free water commonly is very rare or very deep. The soils are commonly coarse-textured and have very high hydraulic conductivity or are very shallow.
Somewhat excessively drained (SED)	>120 cm and most often highly permeable sand, but not quite a permeable as the ED class	Water is removed from the soil rapidly. Internal free water occurrence commonly is very rare or very deep. The soils are commonly coarse-textured and have high saturated hydraulic conductivity or are very shallow.
Well-drained (WD)	>120 cm	Water is removed from the soil readily but not necessarily rapidly. Internal free water occurrence commonly is deep or very deep; annual duration is not specified. Water is available to plants throughout most of the growing season in humid regions. Wetness does not inhibit the growth of roots for significant periods during most growing seasons.
Moderately well-drained (MWD)	60–120 cm	Water is removed from the soil somewhat slowly during some periods of the year. Internal free water occurrence commonly is moderately deep and may persist for short to long periods. The soils are wet for only a short time within the rooting depth during the growing season, but long enough that plants intolerant of saturation are limited. They commonly have a moderately low or lower saturated hydraulic conductivity in a layer within the upper 200 cm.
Somewhat poorly drained (SWPD)	30–60 cm	Water is removed slowly so that the water table is shallow to moderately deep for significant periods during the growing season. Saturation restricts the growth of wetness intolerant plants, unless artificial drainage is provided. The soils commonly have one or more of the following characteristics: low or very low saturated hydraulic conductivity, a high water table, or additional water from external sources–seepage, flooding etc.
Poorly drained (PD)	<30 cm	Water is removed slowly so that the water table is shallow to very shallow during the growing season or the soil remains wet for long periods. Saturation is at or near the surface long enough during the growing season so that most wetness intolerant plants cannot be grown unless the soil is artificially drained. The zone of saturation may not be present continuously directly below depth of plowing. The soils often have low or very low saturated hydraulic conductivity, a low-lying land space position or flat slope.
Very poorly drained (VPD)	<30 cm and >25 cm black surface	Water is removed so slowly that the water table remains at or very near the ground surface during much of the growing season or year. Saturation is very shallow and persists for very long periods or is permanent. Unless the soil is artificially drained, most wetness intolerant plants cannot grow. The soils are commonly level or depressed and frequently ponded.

classes have little to do with the soil's overall permeability or infiltration rates. (The other form of drainage is *external drainage*, which is related to slope and landscape position and will be discussed later in this chapter.)

There are seven classes of agricultural soil drainage. Soils may be excessively drained, somewhat excessively drained, well drained, moderately well drained, somewhat poorly drained, poorly drained, and very poorly drained. The exact definition of what constitutes each drainage class differs regionally. Within each class the depth to the water table ranges from "very deep" (more than 150 centimeters) to "very shallow" (25 cm or less) (table 2–12).

Soils with very shallow or shallow water tables are common in wetlands. These wetland soils are classified as hydric soils. Specifically, hydric soils form when saturation, flooding, or ponding lasts long enough during the growing season to create anaerobic conditions in the soil's upper 30 centimeters (12 inches).

PERMEABILITY

Water in the soil is not static or stationary; it moves. *Permeability* refers to the movement of air and water within the soil and is influenced by soil texture, structure, bulk density, and the type and connectivity of *macropores*. Macropores are the large pores common in sands, or the cracks between the peds in finer textured soil. They also refer to worm and root channels. *Micropores*, on the other hand, are found between the mineral grains within a structural unit or ped.

In sandy soils water rapidly moves downward since the micropores are relatively large and continuous. Clayey soils, on the other hand, have slower permeability, as the micropores are often small and discontinuous. The swelling of clay further lowers the rate of water flow since the size of micropores decreases as the clay particles swell. If a soil has enough expansive clay it may swell enough to close these pores and prevent water flow entirely.

Structure also influences a soil's permeability because of the macropores associated with the structural aggregates (see Box above). Granular structure has a large number of interconnected macropores that readily permit water movement. In platy structure, water flow is more tortuous and thus slower. Horizons with prismatic structure tend to have high amounts of clay; if expansive the clay will swell and restrict flow. *Massive* soils have no macropores, so flow in these soils is entirely related to the characteristics of pores associated with the soil texture.

Permeability is often referred to as *hydraulic conductivity*, which is the amount of water that can move downward through a unit area of soil in a unit of time. *Saturated hydraulic conductivity* is a specific measurement of water movement through soil that is saturated.

INFILTRATION

Unlike permeability, which refers to movement within the soil, infiltration refers to the downward movement of water into the surface of the soil. As with permeability, infiltration is influenced by the soil's texture and other physical properties (structure, bulk density, and the type and connectivity of macropores) as well as by the amount of organic matter in the surface horizons.

A high infiltration rate resists erosion since runoff is low. A high infiltration rate maximizes the amount of rainfall or irrigation that enters the soil. Adding organic matter can dramatically improve infiltration and reduce erosion. Organic matter keeps the soil surface sheltered from the force of raindrops and improves aggregation. Agronomic practices that use cover crops, crop residue, or mulch improve infiltration. Bare soils often develop a crust (thin zone of poor or no structure) on their surface, which inhibits the ready infiltration of water and can enhance erosion.

SOIL ORGANIC MATTER

A key physical component of soil, the organic matter, begins with the biology of soil (chapter 3). All living organisms create waste products and eventually die (figure 2–14). The organic matter of the soil largely results from the decomposing bodies of microorganisms, plants, and animals, as well as their waste. Well-decomposed organic matter is referred to as humus, a dark brown, porous, spongy material that has a pleasant, earthy smell. Maintaining humus is an important aspect of good soil management (table 2–13). The amount of soil organic matter is controlled

Table 2–13. Ten beneficial effects of organic matter or humus in the soil. Organic matter...

1.	Provides carbon, food and an energy source for microbes and other soil organisms
2.	Enhances structure by stabilizing and holding soil particles together, reducing erosion hazard
3.	Improves storage and transmission of air and water in the soil, aiding plant growth
4.	Stores and supplies nutrients (nitrogen, phosphorus, and sulfur) that are essential for plant and animal growth
5.	Retains nutrients
6.	Makes the soil more resistant to compaction
7.	Improves soil tilth (friable, less sticky, and easier to work)
8.	Sequesters carbon from the atmosphere and other sources
9.	Retains pesticides, heavy metals, and many other pollutants, reducing the negative environmental effects of these substances
10.	Insulates the soil

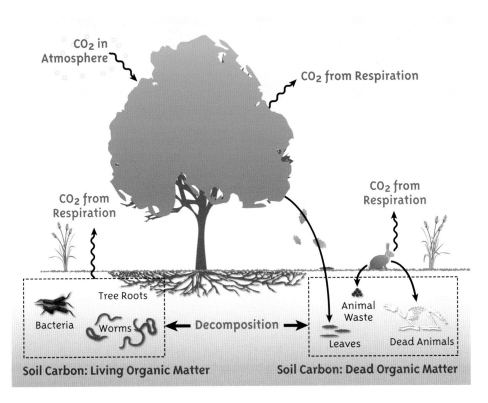

Figure 2–14. The carbon cycle.

by the balance between additions of plant and animal materials and losses through decomposition or erosion, which in turn is controlled by the way soil is managed.

The proportion of total biomass that can become soil organic matter is related to how much biomass is consumed by animals and insects, destroyed by fire, eroded away by water and wind, or harvested for human use. Organic matter decomposes faster when soil is tilled because of the changes in water, aeration, and temperature. Rates of decomposition rise as temperature rises. Anaerobic conditions (without oxygen) tend to preserve organic material as compared with aerobic conditions (with oxygen). Available nutrients (nitrogen in particular) also promote organic matter decomposition.

Soil organic matter improves tilth by reducing crusting, increasing the rate of water infiltration, reducing runoff, and facilitating penetration of plant roots. Organic matter may take decades or centuries to accumulate at the surface of a soil, but it declines rapidly within the first 10 years after a forest is cleared or native grassland is tilled.

SOIL FORMATION

The properties of soil we have just discussed result from the interaction of five general factors. These are the factors of soil formation, known as *CLORPT*: **Cl**imate, **O**rganisms (biological factors), **R**elief or topography, **P**arent materials, and **T**ime or exposure. We'll discuss each of these in detail here.

An understanding of soil formation is valuable for interpreting and managing soils and the land. Soil formation is also called soil genesis. The study of soil genesis means the study of the forces of nature and their effects on the vital skin of Earth.

The factors involved do not act independently—to understand soil genesis one studies current as well as past conditions and their effect on the soil.

CLIMATE

Climate refers to both temperature and rainfall. Climatic conditions bear directly on the weathering of materials in the soil. Weathering is the breakdown or disintegration of rock, minerals, and organic materials at or near Earth's surface, by natural forces. The major climatic forces are temperature and water. Weathering can be subdivided into mechanical (physical) or chemical processes.

Mechanical weathering is the physical breaking of rocks into smaller pieces. This can occur through exfoliation, freeze–thaw cycles, wet–dry cycles, abrasion, and root expansion (figure 2–15). Exfoliation is caused by rapid temperature changes that can expand and crack rocks. Wind may literally sandblast rock surfaces away while water can move particles around, allowing them to hit each other and break apart. Ice can break up rocks by expanding in crevices and cracks, and ice in large amounts (glaciers) can grind rocks into dust if it slides over them. As plant roots grow they may extend into rock cracks, breaking and further cracking the rocks. Although mechanical forces operate slowly, they are effective over long periods of time.

Chemical weathering is the process by which materials decompose due to chemical reactions (figure 2–16, table 2–14). Examples of chemical weathering include carbonation, dissolution, hydration, hydrolysis, and oxidation–reduction reactions. The rate of chemical weathering is influenced by water, oxygen, and the presence of organic and inorganic acids resulting from biochemical activity and temperature. The rate of chemical weathering increases in hot–moist climates, while it is slower in cool–dry climates. The climate in the humid southeastern United States is favorable for a high amount of chemical weathering, unlike the climate of the

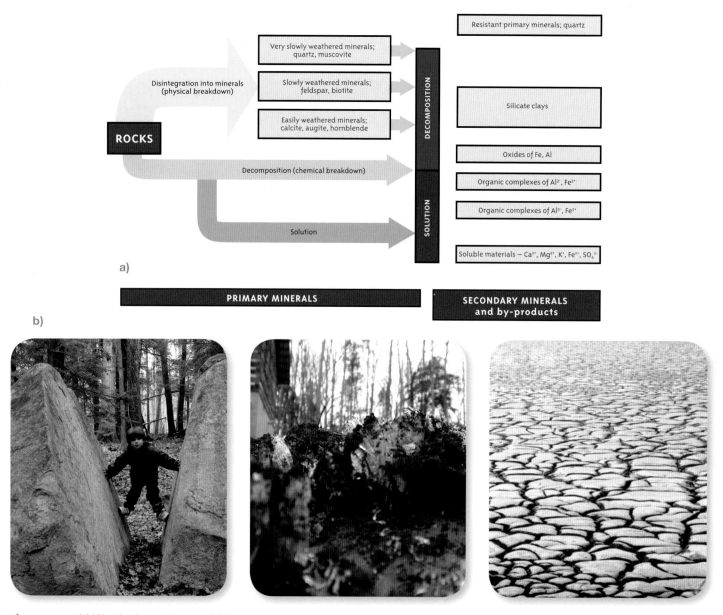

Figure 2–15. (a) Weathering pathways. (b) Temperature changes result in differential expansion of minerals when the temperature is different on the outside vs. the inside. Examples include frost action and mud cracking.

desert Southwest. An example of chemical weathering can be seen in the Parthenon in Athens, Greece. This ancient structure has survived for millennia, but fumes from automobiles and other sources combine with moisture in the air to produce acids that are dissolving this ancient wonder.

Climate is indirectly related to another factor involved in soil formation: organisms. Climate clearly affects the kinds of plants that grow in particular regions, such as cacti in deserts and fir trees in cool climates. Since plants grow in soil they influence the way soil forms.

ORGANISMS OR BIOLOGICAL FACTORS

Microorganisms, plants, and animals help to form soil by adding organic matter, aiding in decomposition, and mixing/aerating (*bioturbation*). Throughout their life cycles all soil organisms contribute to soil

Figure 2-16. Examples of chemical weathering, visible with a microscope and with the naked eye: (from left) feldspar grain weathering to kaolinite in a fluvial derived soil, mica schist saprolite weathering to clay minerals and iron oxides in a residual soil, scanning electron micrograph of a horneblend grain dissolving in glacial till derived soil.

Table 2-14. Chemical weathering examples.

Hydrolysis	$2KAlSi_3O_8 + 2H^+ + H_2O \rightarrow 2K^+ + Al_2Si_2O_5(OH)_4 + 4SiO_2$
	K-feldspar → kaolinite + silica
Hydration	$2Fe_2O_3 + 3H_2O \rightarrow 2Fe_2O_3 \cdot 3H_2O$
Carbon dioxide and carbonic acid	$H_2O + CO_2 \rightarrow H_2CO_3 \rightarrow H^+ + (HCO_3)^-$
	water + carbon dioxide → carbonic acid → H$^+$ ion + bicarbonate ion
	H$^+$ and CO$_2$ in water—acids
Solution	Elements (K, Na, Ca, etc.) released by hydrolysis
	SiO$_2$ moves in solution
	Chemical leached as they are released by other processes
	Organo-metal complex
	Organic complexing agent (ligand)
	Stabilizes complex
Chelation	Weathering can be enhanced—no inhibiting precipitate
	Enhances translocation of metal
	Inhibits decomposition of ligand
	Electrons taken from one substance and given to another
	Electrons come from organic matter as it decomposes
Oxidation/reduction	Oxidation = the production (loss) of electrons
	Reduction = the consumption (gain) of electrons
	$2e^- + 6H^+ + Fe_2O_3 \rightarrow 2Fe(II) + 3H_2O$

formation both actively—through mechanical and chemical weathering, as we will see below—and passively, by adding carbon back to the soil after they die (figure 2–17).

Microorganisms (bacteria, protists, and fungi) are the most common decomposers in the soil (see chapter 3 for more details). Other soil organisms such as earthworms, mites, and nematodes are *detritivores*, which eat detritus. Together these organisms enrich the soil by breaking down organic matter into available nutrients. Apart from releasing nutrients through decomposition (biocycling), they also release organic acids, which break down soil minerals and aid in the overall weathering process.

For discussing soil formation it helps to consider plants in two distinct groups, forest and grasses. In forests a thin O horizon often occurs as a result of falling litter (leaves and limbs). This litter decomposes and accumulates to form an organic layer at the soil surface. The thickness of the O horizon is related in part to climate, as soils in northern forests with cold moist climates have much thicker O horizons than soils in tropical forests with hot moist climates due to a slower rate of microbial decomposition in cold climates. The type of tree litter also has an effect: coniferous litter decomposes more slowly than deciduous litter, so you find thicker O horizons

Figure 2–17. Soil profiles under different vegetation: (from left) mixed hardwood forest, tallgrass prairie, shortgrass prairie, and coniferous forest. The diagrams below further illustrate the effects of vegetation. On the left is a soil developed under coniferous vegetation where organic acids produced as water interacts with the coniferous litter eluviates organic matter, iron, and aluminum from the upper horizons (A and E) to the subsoil (B horizon). On the right is the soil developed under hardwood vegetation where the leaf litter–water interaction does not produce strong organic acids resulting in less dramatic eluviation.

under pine or fir vegetation. This coniferous vegetation is also more acidic, so nutrient leaching (losses) and translocation is enhanced under conifers (figure 2–17).

In grasslands an O horizon is rare, since dead grasses decompose rapidly to build up. Grass roots, on the other hand, die and decompose more slowly, forming a rich, dark A horizon from the constant growth and decay of grass roots (figure 2–17). These soils are common to the Great Plains of North America and the steppes of Europe and Asia. They are among the most fertile soils because of the constant recycling of nutrients.

Animals (including ants and termites) are mostly involved with mixing of the soil, permitting more air and water to enter (figure 2–18). They also mix organic matter throughout the soil and physically bring nutrients translocated deep in the soil back to the root zone.

RELIEF OR TOPOGRAPHY

Relief, or topography, describes the shape, slope, and elevation of the land. Relief has an effect on soil formation through soil wetness, soil temperature, and the erosion rate. The topography of an area can mitigate or enhance weathering caused by climatic factors. It is also responsible for the external drainage condition of a soil or location (figures 2–19 and 2–20). Soils on steep slopes or summits tend to be drier than in flatter areas because of increased runoff. Soils in low-lying areas or flat slopes tend to be wetter since there is little runoff and more water accumulation from upland areas.

Steep slopes have thinner A horizons and overall less soil development. Erosion tends to prevent accumulation of A materials because soil particles are washed away quickly. On gentle slopes with less erosion or removal, thicker A horizons are common. At the "foot" and "toe" of a slope, A horizons tend to be thicker because of deposition from upland areas.

In mountainous areas, the direction the slope faces, also known as aspect, affects temperature and moisture and therefore influences soil organic matter accumulation. In the Northern Hemisphere, steep north- and east-facing slopes have higher amounts of organic matter (thicker A horizons) because they are both cooler and moister than slopes facing south or west (figure 2–21). In deep isolated valleys or coves, A horizons and perhaps O horizons will also be thicker compared with the more exposed upland slopes.

PARENT MATERIALS

Soils have parents just as we do. Soil parent material is the material from which the soil develops. Often the C horizon or R horizon is that material. The majority of the physical and chemical properties of the soil are directly related to the parent material, although these properties may be mitigated by the other factors of formation. The inorganic materials in the soil were ultimately derived from rocks. Not all soils have rocks as their ultimate

Figure 2–18. Larger life forms mix soils as well, including (a) crayfish and (b) squirrels.

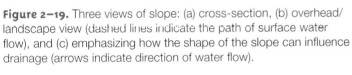

Figure 2–19. Three views of slope: (a) cross-section, (b) overhead/landscape view (dashed lines indicate the path of surface water flow), and (c) emphasizing how the shape of the slope can influence drainage (arrows indicate direction of water flow).

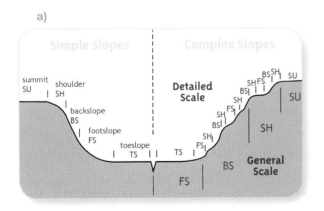

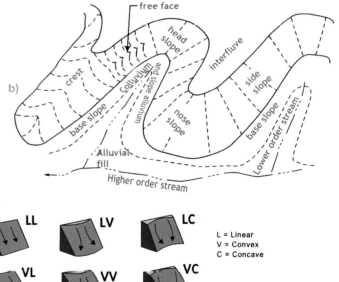

Fig. 2–20. Soils vary with position on a slope (above, from left): summit, shoulder, back-slope, foot-slope, toe-slope, and drainage way.

Figure 2–21. *Aspect*—the direction a slope faces—affects the soil. A north-facing slope is cooler, moister, and has a thicker organic matter layer, while a south-facing is drier, warmer, and has more clay.

North-facing aspect
- Cooler, moister
- More basic cations
- Thicker organic (litter) accumulation
- Thicker A horizon

South-facing aspect
- Dryer, warmer
- More clay
- Stronger argillic horizon
- Thinner A horizon

Table 2–15. Three general categories of parent material.

Organic deposits
Residual material
Saprolite and parent hard rock
Transported
Water
Alluvial (fresh water—fluvial, lacustrine)
Marine (saline to brackish)
Glacial ice (frozen water) and melt water
Gravity
Colluvial deposits (creep, landslides)
Wind
Eolian deposits (loess, eolian sands)
Volcanic (eolian) ash and related pyroclastic deposits

parents; organic soils are derived from decomposed organic (plant and less commonly animal) remains. Parent material is variable but can be broken down into three broad groups: *organic materials*, *residual materials*, and *transported materials* (table 2–15 and figure 2–22).

ORGANIC PARENT MATERIALS

Organic deposits form when the rate of organic matter decomposition is lower than the rate of organic matter accumulation. This occurs in areas that have standing water (anaerobic conditions) or are cold enough to inhibit biological activity, such as Arctic tundra. As plants shed leaves or needles, or die and fall to the surface, any material ending up in a submerged area will decompose slowly if oxygen is low. The organic matter slowly fills in the low-lying wet area, which may be called a swamp, marsh, bog, pocosin, peatland, or fen. While most are in depressional areas, a pocosin (Algonquin name for swamp on a hill) occurs on broad flat regions of the southeastern U.S. coastal plain (see also chapter 7). Drainage may be so slow that the organic surface grows upward. In other cases organic matter can accumulate on slopes where water seeps to the surface, forming localized "hanging bogs." Hanging bogs can be found anywhere groundwater seeps to the surface and decomposing vegetation accumulates, but most commonly in steeply sloping terrain.

RESIDUAL MATERIAL

Bedrock, whatever the type, breaks down into what's known as *residuum* through the weathering process. This residuum is the parent material of the soil and is responsible for the characteristics of the resulting soil profile. Soils that form in place from the underlying bedrock are known as residual soils. Residual soils are common in areas where the surface has been stable for millennia such as the Piedmont of the southeastern United States, parts of South America, parts of Africa, and Australia. Glaciated areas or areas of vast windblown deposits rarely have residual soils present.

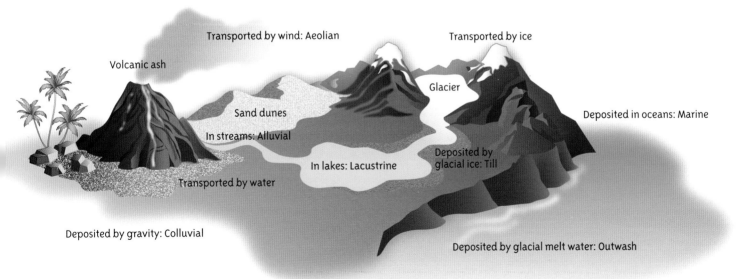

Figure 2–22. The soil landscape is a product of transport and deposition processes. Not all soils have rocks as their ultimate parents; organic soils are derived from decomposed organic (plant and less commonly animal) remains. Parent material is variable but can be broken down into three broad groups: organic materials, residual materials, and transported materials.

Inorganic Parent Materials: Rocks & Minerals

Transported and residual parent materials are composed of rocks and minerals, or inorganic materials. The building blocks of rocks are minerals; thus, minerals are also the building blocks of soils. Minerals are important to soils because they are the most abundant materials that weather to form soil. The amount of decomposed mineral material that remains in the soil affects the availability of many plant nutrients.

A mineral is an inorganic material that has a defined internal crystal structure and chemical composition resulting in specific physical and chemical properties. A rock, on the other hand, is an aggregate composed of one or more minerals. There are a vast number of minerals in the natural world, but only a few are of primary importance to soil formation.

Rocks are divided into three groups: igneous, sedimentary, and metamorphic. The combination of how the rock formed and the minerals it contains has a direct bearing on the physical and chemical properties of soil. In general, rocks rich in mafic minerals (iron and magnesium) are more easily weathered; those rich in felsic minerals (silicon, aluminum, and sodium) are more resistant.

The size of the mineral crystal also affects weathering as rocks with large crystals—such as intrusive igneous rock (gabbro), conglomerate, or schist—are more easily weathered than their fine-grained counterparts of extrusive igneous (basalt), shale, slate, or gneiss. Rocks that have fine crystals (basalt, shale) tend to weather into silt- and clay-rich soils, whereas the coarser textured rocks (granite, sandstone) weather into sandier textured materials.

MINERAL AND RELEVANCE TO THE SOIL

Minerals	Nutrients	Secondary minerals and weathering by-products	Resistance to weathering
Iron oxides $FeOOH$, Fe_2O_3, $Fe(OH)_3$	iron (Fe)	differing Fe minerals	Most
Quartz SiO_2	none	Si in solution	
Muscovite $KAl_2(AlSi_3O_{10})(F,OH)_2$	potassium (K)	clay minerals, Si, Al	
Feldspars $KAlSi_3O_8$, $NaAlSi_3O_8$, $CaAl_2Si_2O_8$	potassium (K), sodium (Na), calcium (Ca),	clay minerals, Si, Al	
Biotite $K(Mg,Fe)_3AlSi_3O_{10}(F,OH)_2$	potassium (K), magnesium (Mg), iron (Fe)	clay minerals, Si, Al	
Amphiboles and pyroxenes $(Mg,Fe)_7Si_8O_{22}(OH)_2$, $(Ca, Mg, Fe)(Si,Al)_2O_6$	calcium (Ca), magnesium (Mg), iron (Fe)	clay minerals, Si, Al	
Calcite, Dolomite, Gypsum $CaCO_3$, $CaMg(CO_3)_2$, $CaSO_4 \times 2H_2O$	calcium (Ca), magnesium (Mg), sulfate (SO_4)	Ca, Mg, SO_4	Least

GOLDICH WEATHERING SEQUENCE

MOST STABLE
- Quartz — SiO_2
- K-Feldspars (microcline and orthoclase) — $KAlSi_3O_8$
- Na-Plagioclase (Albite) — $NaAlSi_3O_8$
- Ca-Plagioclase (Anorthite) — $CaAl_2Si_2O_8$
- Muscovite (mica) — $KAl_2Si_3O_{10}(F,OH)_2$
- Biotite (mica) — $KAl(Mg-Fe)_3Si_3O_{10}(OH)_2$
- Hornblende (double chain amphibole) — $Ca_2Al_2Mg_2Fe_3, Si6O_{22}(OH)_2$
- Augite (single chain pyroxene) — $Ca_2(Al\cdot Fe)_4(Mg\cdot Fe)_4Si_6O_{24}$
- Olivine — $MgFeSiO_4$

LEAST STABLE

BOWENS REACTION SERIES

FIRST TO FORM
- Olivine — $MgFeSiO_4$
- Augite (single chain pyroxene) — $Ca_2(Al\cdot Fe)_4(Mg\cdot Fe)_4Si_6O_{24}$
- Hornblende (double chain amphibole) — $Ca_2Al_2Mg_2Fe_3, Si6O_{22}(OH)_2$
- Biotite (mica) — $KAl(Mg-Fe)_3Si_3O_{10}(OH)_2$
- Muscovite (mica) — $KAl_2Si_3O_{10}(F,OH)_2$
- Ca-Plagioclase (Anorthite) — $CaAl_2Si_2O_8$
- Na-Plagioclase (Albite) — $NaAlSi_3O_8$
- K-Feldspars (microcline and orthoclase) — $KAlSi_3O_8$
- Quartz — SiO_2

LAST TO FORM

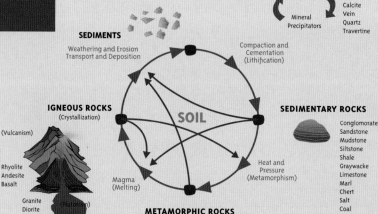

ROCKS BECOME SEDIMENT AND SOIL

SEDIMENTS — Weathering and Erosion, Transport and Deposition

IGNEOUS ROCKS (Crystallization)
- (Vulcanism): Rhyolite, Andesite, Basalt
- (Plutonism): Granite, Diorite, Gabbro

Magma (Melting)

Compaction and Cementation (Lithification)

Heat and Pressure (Metamorphism)

Mineral Dissolution / Mineral Precipitators — Vein Calcite, Vein Quartz, Travertine

SEDIMENTARY ROCKS
- Conglomerate
- Sandstone
- Mudstone
- Siltstone
- Shale
- Graywacke
- Limestone
- Marl
- Chert
- Salt
- Coal

METAMORPHIC ROCKS
- Slate
- Argillite
- Schist
- Gneiss
- Marble
- Metasandstone
- Quartzite
- Greenstone
- Serpentinite
- Chert Breccia

Figure 2–23. Saprolite—low density. It looks like rock but can be easily dug.

Likewise the steep slopes associated with mountain ranges are not stable enough to allow residual soils to develop. Often the C horizons of these soils look very much like the parent rock but are so highly weathered that they can be easily dug into with a shovel (figure 2–23). This highly weathered material is called *saprolite*, which is common in non-glaciated continental piedmonts.

TRANSPORTED PARENT MATERIALS

Both organic and residual parent materials form in place, while some parent materials originated a distance from where they occur now. Transported parent materials are just that: sediments that were transported by water, wind (air), or gravity.

TRANSPORTED BY WATER

Water-transported materials may come from alluvial, marine, or glacial sources.

ALLUVIAL MATERIALS

Alluvial materials (also known as *alluvium*) are transported and deposited by fresh water. The alluvial deposits that commonly become parent soil materials are known as *fluvial* (river-related) and *lacustrine* (lake-related) deposits.

The most common types of alluvial materials are fluvial deposits associated with moving water. A fast-moving mountain stream can carry a large amount of sediment; however, as the sediment reaches the base of the mountain, the slope decreases as does stream velocity, causing the coarsest sediment to be deposited followed by finer sediments. The result is a fan-shaped deposit called an alluvial fan. Alluvial fans are quite common at the foot of mountain slopes in both the Appalachian and Rocky Mountains. Soils within alluvial fans are well drained, and their composition depends on the type of rocks and minerals found on the mountain slopes above.

Streams and rivers commonly overflow their banks and deposit fresh materials on the flood plains. These flood plain soils have poorly developed profiles because of the constant deposition, and most of their character is inherited from the sediment being transported. Often these soils have stratified or layered C horizons as a result of sediment deposition from different flood events. They may also contain buried soils.

Rivers and streams are dynamic, often cutting downward or incising into the underlying material. The result is a new flood plain topographically lower than the previous flood plain. The higher exposed area (old flood plain) is referred to as a terrace. As with flood plain soils, the parent material on a terrace is water-transported alluvium, but as these soils are older they have more developed soil profiles.

Whenever a river or stream flows into a still body of water such as a lake, any suspended sediment settles out and is deposited on the lake bottom as lacustrine materials. These deposits can become thick over time and eventually may fill up the lake. The sediments in the lake bottom are often layered with a winter (fine or clayey texture) and summer (sand or silt texture) layer called *varves*. Over time lake levels can fluctuate and expose lacustrine deposits to the processes of soil formation. Some of the largest areas of these soils occur on the shores of the Great Lakes, in the Southwest (pluvial lakes), and in the glaciated areas of the northern plains (Lake Agassiz).

MARINE MATERIALS

Sediments deposited in the ocean or estuaries are marine materials. Once sediments reach the ocean they are sorted by soil particle size, with larger particles deposited close to the shore or in high energy areas like beaches. The smaller particle clays get deposited farther out or in low energy areas like tidal marshes. The sediments may be reworked to form barrier islands and tidal marshes, such as the Outer Banks of North Carolina. As sea level changes, these sediments may be exposed or inundated. Along the East Coast and across the Gulf of Mexico exposed marine sediments have formed a flat, broad coastal plain. These coastal plains are often older than the alluvial materials that dissect or cross them and thave thick, well-developed horizons.

GLACIAL MATERIALS

Continental glaciers covered much of the northern portion of the Northern Hemisphere over the last few million years. These massive sheets of ice, often as much as two miles thick, removed existing soils and ground up rocks as they moved south. Once the glaciers retreated, the mixed-up, ground-up transported rocks and soil remained to become parent material for the region's soils. This material is

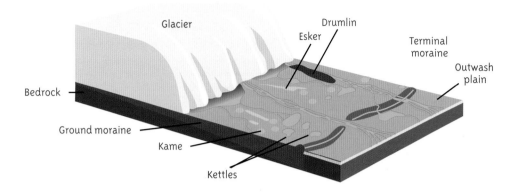

Figure 2–24. Glacial materials: Till—ground moraine, terminal moraine, drumlin. Outwash—eskers, kame, outwash plain. Kettles are ponds where glacial ice was left behind during the ice retreat. When they melted, the depressions became ponds.

collectively called glacial drift, of which there are two types: glacial outwash and glacial till (figure 2–24).

Glacial outwash is the material washed away from the glacier by melt water and is also referred to as *glaciofluvial material*. It is often sandy, contains a lot of well-rounded gravels, and is often sorted or layered. *Glacial till* is the material deposited directly from the glacial ice. It is commonly finer than glaciofluvial materials and poorly sorted and rarely layered. Stones and boulders are common and are rarely rounded. Because of the way it was deposited it can be dense and often restricts water movement and plant roots. The glacial deposits are common in northern parts of the North America, Europe, and Asia.

TRANSPORTED BY WIND

Materials that were deposited by wind are known as *eolian* material. Eolian deposits are separated in part by the particle size of the material. *Loess* is generally considered to be dominantly silt sized, whereas dune or eolian sands are dominantly sand sized. Volcanic ash is another type of eolian material that is carried by wind.

Loess consists of windblown silts that originated in broad flood plains and were carried upland by wind. Loess-derived soils are highly productive agricultural soils, but are highly susceptible to erosion. Loess soil is most common in the midwestern and Great Plains area of North America and in the loess plateau of central Asia.

Strong winds can move and deposit sand into sand dunes such as the Sahara Desert. Unless they are well vegetated, sand dunes are unstable and are reshaped in response to prevailing winds. Not all eolian sands are in sand dunes. Areas of New England, the southwestern plains and the southeastern coastal plain are blanketed by very fine to medium sand (from 30 centimeters, or 12 inches, to a meter or more thick). Particularly in the southwestern plains, these soils require irrigation and good management to prevent wind and water erosion (see chapter 6).

Volcanic eruptions can expel large amounts of lava or volcanic ash. Soils developed on lava are considered to have a residual parent material, whereas soils developed from volcanic ash are considered to have wind-transported parent material. Wind can carry volcanic ash and cinders over large areas. The Cascade Range of the northwestern United States and much of Japan are examples of this type of volcanic deposit. Volcanic ash has unique properties including a high phosphorus-fixing ability, which means these soils chemically bind large amounts of phosphorus, and make it unavailable for plant use (see chapter 4). They can also hold a great deal of water, which makes them unstable on slopes and for building.

TRANSPORTED BY GRAVITY

The last group of transported materials is that moved by gravity, either as soil creep or as landslides. This material is referred to as *colluvium*, or colluvial deposits. In mountains with long steep slopes, surface material (soil, organic materials, and rock fragments) may move downhill under the influence of gravity. Often this movement is initiated by water saturating the sediment and acting as a lubricant, but the water is not the primary reason the sediment moves. The material is disorganized and eventually accumulates on the lower portion of slopes or in depressions. Rock fragments found in colluvium are angular in shape and the overall material is unsorted.

TIME

The final factor involved in soil formation is time. The length of time that a surface has been exposed to weathering and biotic forces affects the overall nature and properties of the soil. Soil may range in age from days to many millennia. **It takes hundreds to thousands of years for these factors of formation to create one inch of soil from parent material** (figure 2–25, table 2–16).

Soil formation is a continuous series of processes. Recently deposited materials, whether from a flood or landslide or wind-blown dunes at a beach, exhibit no features (horizons) from soil formation processes. The formation time clock for these soils in effect has been reset. In contrast, terraces that occur above an active floodplain show more soil development by way of more soil horizons.

As soils age they develop thicker A and B horizons. Young soils or poorly developed soils often look like the parent materials they are forming from and may only have a thin A horizon directly over a C horizon. Older, highly weathered or well-

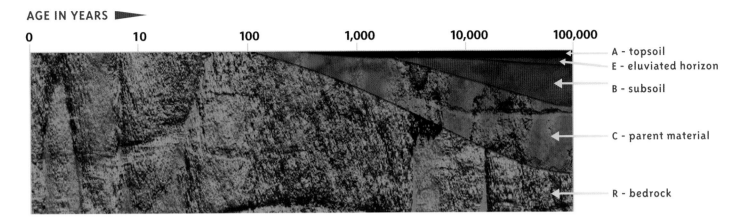

Figure 2–25. Time and soil thickness. Soil horizons take hundreds to thousands of years to form.

Table 2–16. In the time it took an inch of top soil to form the following happened...

IN THE TIME IT TOOK TO FORM ONE INCH OF SOIL...

World population exceeds 7 billion	2011
Barack Obama elected President	2008
Terrorist attacks on the World Trade Center and Pentagon	2001
New planets that could sustain life are discovered orbiting a neighborhood star	1996
The Berlin Wall is torn down, signaling an end to the Cold War	1989
Viking I becomes the first spaceship to reach Mars	1976
U.S. Armed Forces withdraw from Vietnam	1975
Congress passes Endangered Species Act	1973
Richard Leakey discovers hominid skull, 2.6 million years old in northern Kenya	1972
First "Earth Day" is observed	1970
American astronaut Neil Armstrong walks on the moon	1969
Thurgood Marshall becomes first African American member of the U.S. Supreme Court	1967
Rev. Martin Luther King, Jr. receives the Nobel Peace Prize in Oslo, Norway	1964
President John F. Kennedy is assassinated in Dallas, Texas	1963
Soviet cosmonaut Yuri Gagarin becomes first human to orbit Earth	1961
Korean War ends	1953
Dr. Jonas Salk successfully tests polio vaccine	1952
Mao Tse Tung rises to power in China	1949
The transistor is invented in Bell Laboratories in the United States	1948
ENIAC, the first successful electronic digital computer, becomes operational	1946
World War II ends	1945
Child film actress Judy Garland stars in *The Wizard of Oz*	1939

That inch of soil didn't begin forming so recently! Keep reading...

IN THE TIME IT TOOK TO FORM ONE INCH OF SOIL...

A Coelacanth, a fish thought extinct for 65 million years, is caught of the coast of Africa	1938
First local conservation district is created	1937
Drought leads to severe dust storms in "Dust Bowl" region of Great Plains	1934
Frances Perkins, Secretary of Labor, becomes first woman in U.S. cabinet	1933
Hirohito becomes 124th Emperor of Japan	1926
19th amendment to the U.S. Constitution gives women the right to vote	1920
World War I ends	1918
Russian Revolution begins	1917
Jeanette Rankin (R—Montana) elected first U.S. Congresswoman	1916
Albert Einstein formulates his general theory of relativity	1915
First Model "T" Ford rolls off assembly line	1908
Upton Sinclair's novel, *The Jungle*, leads to U.S. Pure Food and Drug Act	1906
Orville Wright makes first successful flight in self-propelled airplane	1903
Gugliemo Marconi tests radio transmissions between England and Newfoundland	1901
Dr. George Washington Carver begins to develop over 300 products from the peanut and 118 from the sweet potato	1896
German Physicist Wilhelm C. Roentgen discovers X-rays	1895
San Francisco Examiner begins printing first newspaper comic strip	1892
Thomas Edison invents phonograph	1877
Alexander Graham Bell invents telephone	1875
U.S. experiences worst grasshopper plague in its history	1874
Congress founds first national park, Yellowstone	1872
U.S. Civil War ends & 13th Amendment to the U.S. Constitution abolishes slavery	1865
President Abraham Lincoln delivers Gettysburg Address	1863
Gregor Mendel begins formal experiments in genetics	1856
American writer and naturalist Henry David Thoreau publishes *Walden*	1854
Frederick Douglass establishes the abolitionist newspaper, *The North Star*	1847
Samuel F.B. Morse develops code for electric telegraph system	1838
Texas proclaims independence from Mexico	1836
Louis Braille publishes system of writing for the blind	1829
President James Monroe formulates doctrine of U.S. policy opposing outside interference in the Americas	1823
Construction of Erie Canal begins	1817
Kamehameha I becomes ruler of Hawaii and establishes Kamehameha dynasty	1810
Ludwig van Beethoven premieres Fifth and Sixth Symphonies	1808
Lewis and Clark begin expedition to the American Northwest	1804
English physician Edward Jenner develops vaccination against smallpox	1796
Eli Whitney invents cotton gin	1793
Bill of Rights is ratified	1791
French Revolution begins with an attack on the Bastille	1789
The U.S. Constitution is ratified	1788
American Revolutionary War ends	1781

IN THE TIME IT TOOK TO FORM ONE INCH OF SOIL...

Continental Congress adopts Declaration of Independence	1776
American Colonists throw British tea into Boston Harbor	1773
Crispic Attucks becomes the first person and African American to die in the Revolutionary War in the "Boston Massacre"	1770
French and Indian War ends	1763
Thomas Jefferson is born	1743
George Washington is born	1732
First permanent Native American school is created in Williamsburg, Virginia	1720
Italian musician Bartolommeo Cristofori invents the piano	1709
Benjamin Franklin is born	1706
First public school is established in America	1689
Edmund Halley observes Great Comet, which is later named for him	1682
The Dodo, a large flightless bird, becomes extinct	1681
First book is printed in Native American language	1653
First newspaper, *Gazette de France*, is published in Paris	1631
Pilgrim leaders at Plymouth Colony establish a governing authority through Mayflower Compact	1620
Native Americans teach colonists how to raise corn	1608
John Smith founds first permanent English colony in North America at Jamestown, Virginia	1607
Spain founds colony in St. Augustine, Florida	1565
Shakespeare and Galileo are born	1564
Juan Ponce de Leon discovers Florida	1513
Michelangelo finishes painting Sistine Chapel ceiling	1512
Leonardo da Vinci completes *Mona Lisa*	1505
The Incan Empire reaches its height	1500
The toothbrush is invented by a Chinese dentist	1498

developed soils may be less fertile than young soils because of more nutrient loss from leaching and erosion.

All the soil-forming factors combined affect all soils, even those that appear to be unchanging. The more general soil processes discussed in chapter 1—additions, losses, transformations, and translocations—are ongoing but are mitigated or controlled by the factors of formation discussed here.

SUMMARY

After reading through this chapter you should begin to understand how complex a handful of soil is. That handful is composed of sand, silt, and clay particles as well as organic matter. The properties of soil go far beyond its texture, however. The texture (percentage of sand, silt, and clay) and the associated pore space control many of the other physical properties of the soil—bulk density, water holding, permeability, and infiltration, to name a few. Of all the soil particles, clay is the smallest but is perhaps the most important and is likely formed directly in the soil from weathering.

Although texture is important, how those particles are arranged into a soil's structure also influences soil properties. A well-structured soil will have more permeability, have better infiltration, and be less erodible than a poorly structured or unstructured soil. Likewise, soil consistence, which is related not only to the amount of clay but also to the type of clay minerals present, can help us identify soils with shrink–swell clays that may pose a challenge for soil management.

Since soil does not exist just as a single handful we have also examined the soil profile and soil color. By doing so we begin to use the language of soil with O, A, E, B, C, R master horizons, the multitude of subhorizons, and all the combinations thereof. Horizons are often first identified by color changes, and the Munsell color system shows that color has three dimensions: hue, value, and chroma. Color is also used to identify redoximorphic features. These features and the colors help us to identify the depth to seasonal saturation in the soil as well as the aerobic or anaerobic nature of the soil.

Soil water is important not just for the depth to the water table or saturation but also for its movement. Infiltration is the movement of water into the soil from the surface, while permeability is the movement of water through the soil. Soil water flows by gravity or by capillary attraction (adhesion and cohesion).

In a broader sense, we can discuss water in the soil based on internal and external drainage. Internal drainage relates to how the water moves in the soil, while external relates to how it moves on the landscape. Agricultural drainage classes do not relate to water movement at all but to the depth to the water table or saturation during the growing season. (This is not to be confused with the depth to free water or saturation at any time during the year.)

Finally, we took a big step back and looked at how soil develops its morphology and other soil properties. The five factors of soil formation—climate, organisms, relief, parent material, and time, also known as CLORPT—work together on any given soil to mold it into what we see today. Because CLORPT factors do change, the soil morphology represents how those factors interacted in the past. Soils can be a window to both current conditions as well as to the past. These factors will continue to guide our understanding of soils in subsequent chapters.

ADDITIONAL READING

Brady, N.C., and R.R. Weil, 2008. *The Nature and Properties of Soils.* 14th ed. Pearson Education, Inc., Upper Saddle River, NJ.

Foth, H.D. 1990. *Fundamentals of Soil Science.* 8th ed. John Wiley & Son, Inc., Hoboken, NJ.

Kohner, H., and D.P. Franzmeier, 1995. *Soil Science Simplified.* 4th ed. Waveland Press Inc., Long Grove, IL.

CREDITS

2–1, D. Lindbo; 2–2, D. Lindbo, John Kelley; 2–3, John Kelley; 2–4, D. Lindbo; 2–6, USDA-NRCS; 2–7, D. Lindbo, B. Richards; 2–10, D. Lindbo; 2–11, D. Lindbo; 2–15, D. Lindbo; drawing after Brady and Weil, 2008; 2–16, D. Lindbo; 2–17, drawing, James Nardi; photos, USDA-NRCS and D. Lindbo; 2-18, D. Lindbo; 2-20, D. Lindbo; 2–21, USDA-NRCS; 2–23, D. Lindbo. Chapter opener image, iStock. Table 2–5, data from H.D. Foth, 1978; Table 2-8, data Brady and Weil, 2008; Table 2–16 excerpted from http://soil.gsfc.nasa.gov.

Copyright © 2012. Soil Science Society of America, 5585 Guilford Rd., Madison, WI 53711-5801, USA. *Know Soil, Know Life.* David Lindbo, Deb A. Kozlowski, and Clay Robinson, Editors
doi:10.2136/2012.knowsoil.c2

GLOSSARY

AGGREGATE A group of primary soil particles that cohere to each other more strongly than to other surrounding particles.

AGRICULTURAL DRAINAGE CLASS Classes used to describe the depth to the water table during the growing season. The exact depth varies from state to state. Agricultural drainage classes have little to do with the soil's overall permeability or infiltration rates.

ALLUVIUM Sediments deposited by running water of streams and rivers. It may occur on terraces well above present streams, on the present flood plains or deltas, or as a fan at the base of a slope.

ASPECT The compass direction that the slops faces in relation to the sun.

BIOTURBATION Mixing of the soil by organisms—worms, ants, moles, tree roots.

BULK DENSITY The mass of dry soil per unit bulk volume. The value is expressed as megrams per cubic meter, Mg m^{-3}.

CAPILLARITY Combination of the attractive forces between a water molecule and a surface (adhesion) and the attractive forces between water molecules (cohesion).

CAPILLARY WATER The water held in the "capillary" or small pores of a soil.

CAPILLARY ZONE, OR CAPILLARY FRINGE Zone, just above the water table or zone of saturation, where water is held in the soil against the force of gravity by the capillary attraction of water to the soil particles. Water content in the zone is highest near the water table and decreases with distance above the water table.

CHROMA The relative purity, strength, or saturation of a color; directly related to the dominance of the determining wavelength of the light and inversely related to grayness; one of the three variables of color. See also Munsell color system, hue, and value.

CLORPT Acronym to help students remember the factors of soil formation: CLimate, Organisms, Relief, Parent material, and Time.

COLLUVIUM Unconsolidated, unsorted earth material being transported or deposited on sideslopes and/or at the base of slopes by mass movement (e.g., direct gravitational action) and by local, unconcentrated runoff.

CONSISTENCE The attributes of soil material as expressed in degree of cohesion and adhesion or in resistance to deformation or rupture.

DETRITIVORES Organisms that eat detritus (organic matter) that is added to the soil as organisms die .

ENDOSATURATION The soil is saturated with water in all layers from the upper boundary of saturation to a depth of 200 cm or more from the mineral soil surface. See also episaturation.

EOLIAN Pertaining to earth material transported and deposited by the wind including dune sands, sand sheets, loess, and parna.

EPISATURATION The soil is saturated with water in one or more layers within 200 cm of the mineral soil surface and also has one or more unsaturated layers with an upper boundary above 200 cm depth, below the saturated layer(s) (a perched water table). See also endosaturation.

EVAPOTRANSPIRATION The combined loss of water due evaporation from the ground surface and transpiration form plants.

EXTERNAL DRAINAGE Related to how water moves on the landscape due to slope and landscape position.

Fluvial Deposited by rivers.

GLACIAL TILL Unsorted sediments deposited directly form glacial ice containing a wide range of particle sizes and rock fragments. Glacial till often has a high bulk density.

GLACIOFLUVIAL MATERIAL Material moved by glaciers and subsequently sorted and deposited by streams flowing from the melting ice. The deposits are stratified and may occur in the form of outwash plains, deltas, kames, eskers, and kame terraces.

GLEY Developed under conditions of poor drainage resulting in reduction of iron and other elements and in gray colors and mottles. (Not used in current U.S. system of soil taxonomy; also, a soil condition resulting from prolonged soil saturation, which is manifested by the presence of bluish or greenish colors through the soil mass or in mottles (spots or streaks) among the colors. Gleying occurs under reducing conditions, by which iron is reduced predominantly to the ferrous state.

GRAVITATIONAL WATER Water that moves into, through, or out of the soil under the influence of gravity.

GROUNDWATER The zone where all of the pores in the soil are filled with water and water will flow due to gravity. The water table refers to the top of this zone.

HORIZON A layer of soil or soil material approximately parallel to the land surface and differing from adjacent genetically related layers in physical, chemical, and biological properties or characteristics, such as color, structure, texture, consistency, kinds and number of organisms present, degree of acidity or alkalinity, etc.

HORIZON, MASTER Major soil horizons designated by a capital letter. Each master horizon has a single dominant characteristic i.e. O horizon is dominated by organic matter, C horizon is parent material.

HORIZON, SUBORDINATE Horizons that have a specific property that allows them to be further subdivided.

HUE A measure of the chromatic composition of light that reaches the eye; one of the three variables of color. See also Munsell color system, chroma, and value, color.

HYDRAULIC CONDUCTIVITY Amount of water (liquid) that can move through a unit area of soil in a unit of time (see permeability).

INFILTRATION The entry of water into soil.

INTERNAL DRAINAGE Related to how water moves through the soil due texture, structure and consistence.

ISOMORPHIC SUBSTITUTION The replacement of one element for another in a mineral structure or crystal without disrupting the mineral.

LACUSTRINE Deposited in lakes.

LOESS Material transported and deposited by wind and consisting of predominantly silt-sized particles.

MACROPORES Large pores responsible for preferntial flow and rapid, far-reaching transport.

MASSIVE A structureless soil condition where there are no aggregates and appears cohesive. No discernible planes of weakness exist in massive material.

MASSIVE-ROCK CONTROLLED FABRIC A structureless soil condition where the material has retained its rock fabric but has been weathered to the point where it can be easily excavated. Material often breaks apart along planes of weakness related to the original rock fabric .

MATRIX COLOR The dominant color of the horizon, or layer.

MICROPORES Pores that are sufficiently small that water within these pores is considered immobile, but available for plant extraction, and solute transport is by diffusion only.

MUNSELL COLOR SYSTEM A color designation system that specifies the relative degrees of the three simple variables of color: hue, value, and chroma. For example: 10YR 6/4 is a color (of soil) with a hue = 10YR, value = 6, and chroma = 4. See also chroma, hue, value, color.

ORGANIC MATERIALS Soil materials that are saturated with water and have 174 g kg^{-1} or more organic carbon if the mineral fraction has 500 g kg^{-1} or more clay, or 116 g kg^{-1} organic carbon if the mineral fraction has no clay, or has proportional intermediate contents, or if never saturated with water, have 203 g kg^{-1} or more organic carbon.

PED A unit of soil structure such as a block, column, granule, plate, or prism, formed by natural processes (in contrast with a clod, which is formed artificially).

PERMEABILITY (i) The ease with which gases, liquids, or plant roots penetrate or pass through a bulk mass of soil or a layer of soil. Since different soil horizons vary in permeability, the particular horizon under question should be designated. (ii) The property of a porous medium itself that expresses the ease with which gases, liquids, or other substances can flow through it, and is the same as intrinsic permeability k.

POROSITY The volume of pores in a soil sample (nonsolid volume) divided by the bulk volume of the sample.

REDOXIMORPHIC FEATURES Redoximorphic concentrations, redoximorphic depletions, reduced matrices, and other features indicating the chemical reduction and oxidation of iron and manganese compounds resulting from saturation.

REDOXIMORPHIC Morphology related to the reduction, oxidation and translocation of carbon, iron, manganese and sulfur.

RESIDUAL MATERIALS Unconsolidated and partly weathered mineral materials accumulated by disintegration of consolidated rock in place.

RESIDUUM Unconsolidated, weathered, or partly weathered mineral material that accumulates by disintegration of bedrock in place.

RIPARIAN ZONE The above (vegetation) and below ground (soil) area adjacent to a water body typically a stream or river.

SAPROLITE Chemically weathered rock that has not lost its original rock fabric or changed its volume but has lost some of its original mass—in other words it looks like rock but has a lower bulk density than unweathered rock and can be dug with a shovel.

SATURATED HYDRAULIC CONDUCTIVTY Amount of water (liquid) that can move through a unit area of soil in a unit of time when the soil is saturated.

SATURATION All the soil pores are filled with water.

SINGLE-GRAINED A structureless soil condition where the particles are not aggregated. Non-cohesive materials.

STRUCTURE The combination or arrangement of primary soil particles into secondary units or peds. The secondary units are characterized on the basis of size, shape, and grade (degree of distinctness).

STRUCTURELESS No observable aggregation or no definite and orderly arrangement of natural lines of weakness. Massive, if coherent; single-grain, if noncoherent.

TEXTURE The relative proportions of the various soil size separates in a soil.

TOPSOIL (i) The layer of soil moved in cultivation. Frequently designated as the Ap layer or Ap horizon. See also surface soil. (ii) Presumably fertile soil material used to topdress roadbanks, gardens, and lawns.

TRANSPORTED MATERIALS Parent materials that have been moved and deposit some distance from their point of origin.

VADOSE ZONE Zone of unsaturated soil above the water table.

VALUE Relating to color, the degree of lightness or darkness of a color in relation to a neutral gray scale. On a neutral gray scale, value extends from pure black to pure white; one of the three variables of color.

VARVE(S) Layered lacustrine mineral sediment the are annually deposited with winter (fine or clayey texture) and summer (sand or silt texture) layer similar to the annual growth rings in a tree.

WATER-HOLDING CAPACITY The amount of water that a given material can hold after gravitation al water has been removed.

WATER TABLE Top of the zone where all the pore in the soil are filled with water and water will flow due to gravity.

CHAPTER 3

SOIL BIOLOGY: THE LIVING COMPONENT OF SOIL

Soil is literally full of life. It is often said that a handful of soil has more living organisms than there are people on planet Earth. For this to be true, it means the size of each organism must be exceedingly small (figure 3–1). In fact, the majority of soil organisms are so small that one needs a microscope just to observe their basic shape. Although small, the activity of these organisms is vital for life as we know it. The organisms are active in decomposing plant residues, animals and their waste products, and industrial wastes; recycling carbon; providing nutrients to plants; forming soil pores to provide air and water for plant growth; and maintaining a sustainable environment in soil.

On first observation, soil may appear to be a rather inert material on which we walk, build roads, construct buildings, and grow plants. On closer observation, however, we observe that soil is an environment literally swarming with living organisms, including microbes—archaea, bacteria, other single-cell organisms, and fungi—and a wide variety of larger invertebrate and vertebrate animals (fauna). The fauna includes midsize springtails, mites, nematodes, worms, ants, and insects that spend all or part of their life underground, and larger organisms such as burrowing rodents. All of these organisms are important in making up the environment we call soil and in bringing about numerous transformations that are vitally important to life both above and below ground.

Soil is a complex habitat consisting of mineral and organic particles; living organisms including plant roots, microbes, and larger animals; and pores filled with air or water (based on recent rains). These properties change in space and time. Not only are there large and small soil mineral particles (sand, silt, and clay), some coated with organic matter, but also large and small pore spaces and actively growing and decaying plant roots.

In addition, soil receives many carbon-based materials. Some of these are natural plant and animal wastes, but some come from human activities and include sewage sludge, landfill materials, pesticides, and oily wastes. **Soil acts as a biological incinerator: soil organisms decompose these materials, allowing growth of more soil organisms** (figure 3–2). If waste compounds did not degrade, they would accumulate in our environment over eons of time. For example, if the chitin of insect exoskeletons did not break down, we would be knee-deep with insect body parts. Lucky for us, organisms in the soil can decompose the chitin. ▷▬▶

CHAPTER AUTHOR

TOM LOYNACHAN

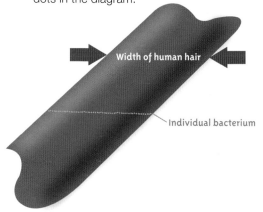

Figure 3–1. Why can't I see soil microorganisms? A human hair may have a diameter of about 50 μm (micrometers), whereas a single bacterium may have a diameter of 1 μm—see the tiny yellow dots in the diagram.

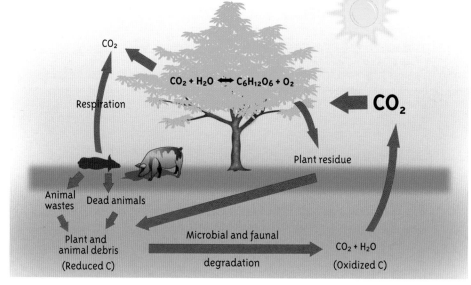

Figure 3–2. Soil life largely depends on the carbon cycle. Plants capture sunlight energy and reduce the carbon from carbon dioxide into plant components that serve as a food source. Much of soil life oxidizes this carbon back to carbon dioxide.

Experts estimate that planet Earth is about 4.5 billion years old, and fossil records indicate that life has existed in its rocks and soils for about 3.8 billion years. Early life consisted of single-cell forms living in an *anaerobic* (without oxygen) environment. For comparison, scientists believe modern humans originated in Africa about 200,000 years ago. A sobering statistic: for every day humans have existed on Earth, bacteria have been here approximately 52 years! Soil organisms have had lots of time to adapt to the environments in which they live, and this probably helps explain the huge diversity of organisms in soils.

Much of the diversity can be explained by the ways these organisms obtain energy. Some organisms can use sunlight energy as plants do; others oxidize reduced carbon for their energy, as discussed to the right, much as we do when we eat a piece of bread. Still other organisms get their energy entirely from inorganic transformations. For example, nitrifying bacteria need only inorganic compounds to generate everything they need for life. They get their energy (electrons) from the oxidation of ammonium (NH_4^+) in the process called *nitrification*. For every mole of ammonium transformed to *nitrate* (NO_3^-), eight moles of electrons are released.

$$NH_4^+ + 2O_2 \leftrightarrow NO_3^- + H_2O + 2H^+ + electrons$$

Other bacteria can use iron or sulfur compounds for their energy.

Organisms must be able to survive and compete in the soil environment. Those organisms unable to adapt to soil conditions will die. It is not necessarily the conditions of the bulk soil that the organisms must adapt to but conditions within a small area (microsite). Important conditions that may vary include pH, moisture, temperature, redox potential (oxygen level), mineral nutrition, the presence or absence of growth factors (organic compounds some organisms require), and positive or negative interactions with other organisms. Negative interactions with other organisms include predation (figure 3–3).

Electrons as Energy: Oxidation and Reduction

During decomposition, microorganisms benefit by the energy released. They use part of this energy to make new body tissue during reproduction and part of the energy to fuel their many activities. Energy levels often are the limiting factor in the growth of organisms in soil. The energy released comes from a series of complex biological reactions revolving around oxidation and reduction. *Oxidation* is the *loss* of electrons, and *reduction* is the *gain* of electrons. If one compound loses a given number of electrons, another compound must gain the same number of electrons. For many organisms (including humans), reduced carbon is their main energy and carbon source. In the carbon cycle, microbes and plants take carbon dioxide from the air and "fix" it by adding electrons, which reduces the carbon to form sugars (or carbohydrates). Plants, animals, and many soil organisms use part of this fixed carbon to make new cells and tissue. The rest of the carbon is oxidized (loses electrons) for metabolic energy, liberating carbon dioxide in the process. The study of the chemistry of living organisms is biochemistry.

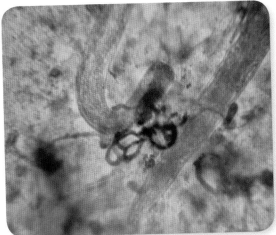

Figure 3–3. Life in the soil can be harsh. Some soil fungi capture nematodes and use them for an energy source. This fungus produces loops that catches the nematode; the fungal strands (hyphae) then penetrate the nematode and digest it (1000× magnification).

ORGANISMS OF THE SOIL

Soil organisms can be classified in several ways: by their size, diet, requirements for oxygen, or activity (what they do). Size often places organisms into one of three categories: micro, meso, or macro. *Microorganisms* are too small to be seen with the unaided eye, normally smaller than 0.1 millimeter (as we learned above, some are a fraction of the width of a human hair). *Mesoorganisms* are middle size (as the name implies) and are 0.1–2 millimeters. The *macroorganisms* are relatively large, greater than 2 millimeters. As a general rule, the smaller the organism, the more numerous are their numbers in soil. Microorganisms in the soil rarely act alone: they normally occur as populations within a complex community (figure 3–4). It is difficult to study the whole community, however, without understanding individual members making up that community.

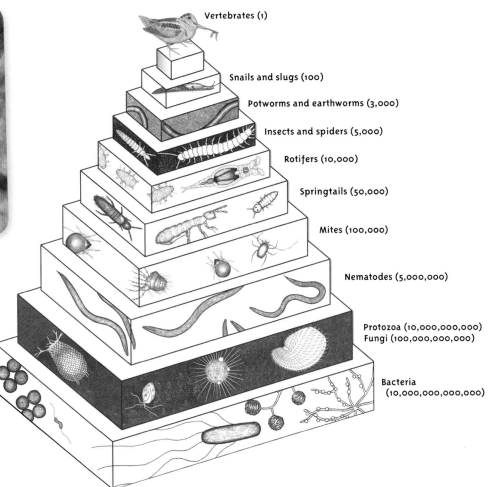

Figure 3–4. The diversity and abundance of living organisms found in most soils (numbers are expected populations per square meter of surface soil). The base of the pyramid has lots of organisms, which tend to be small and support life for organisms above. Upward through each level, fewer organisms remain until you get to the top with a single bird. The smallest organisms, it turns out, are the most abundant.

Counting and Measuring Very Small Things

Because of the large number of microorganisms found in soil, scientists often use *scientific notation* based on powers of the base number 10. Thus, 10,000,000,000 microbes (1 followed by 10 zeros or 10 billion) equals 1×10^{10} (often abbreviated 10^{10}) microbes. Scientists also use the metric system for expressing sizes, which also is based on powers of the base number 10. With small organisms, the measurements are in fractions of a meter: a decimeter is 0.1, centimeter is 0.01, millimeter is 0.001, micrometer is 0.000,001, and nanometer is 0.000,000,001 of a meter. In scientific notation, a nanometer can be expressed as 1×10^{-9} meters. (One meter equals 39.37 inches.)

MICROORGANISMS

VIRUSES are submicroscopic particles that are too small to be viewed with a light microscope. They are among the smallest of soil organisms and some might even argue whether they are living because they are obligate intracellular parasites that can only reproduce inside the living cells of other organisms. The infected host cell provides the metabolic energy needed for generating new viruses, including replicating their nucleic acid (DNA or RNA) and replicating their surrounding protein coats or envelopes. Once the viruses have multiplied, the cells rupture and release the viruses into the environment. Viruses normally are found in numbers as high as 10^{10} per gram of soil. This equals 4.5×10^{12} viruses per pound of soil (if expressed as dollars with one hundred pennies equaling a dollar, this would be $45,000,000,000).

Viruses are known to infect many soil organisms, and many are very specific to the host cells they infect. Soil viruses may alter the *ecology* of soil communities by their ability to transfer genes from host to host or by their ability to kill their host cell. Some of the most intensely studied soil viruses cause disease of higher plants. They may be transmitted by direct contact or may be transmitted through the soil by other organisms such as nematodes. Much yet remains to be discovered about the ecological importance of soil viruses.

The average virus is about one tenth to one hundredth the size of the average bacterium and is usually less than 0.2 micrometers long. The shapes are quite variable, ranging from round to elliptical or rod shaped. Some bacteriophages—viruses that attack bacteria (figure 3–5)—are differentiated into a head and tail portion, with the nucleic acid contained in the head. The tail is the point of attachment to the bacterium through which the viral nucleic acid is inserted into the bacterial cell.

BACTERIA AND ARCHAEA are unicellular organisms, usually 1 to 3 micrometers, that can reproduce independently of other organisms. They are usually the most numerous microorganisms in soil (table 3–1) that can live independent of other organisms, and often exceed 10^7 to 10^8 per gram of soil. Even though they are small and single celled (figure 3–6), they contain all the complexities of life. They consume food, respire, reproduce, and excrete waste products.

In the past, bacteria and archaea were classified together because they have a

Table 3–1. Expected population size† of smaller organisms per gram of a healthy soil.

Organism	Size	Numbers per gram‡
Bacteria	0.2–2 µm	100,000,000
Actinomycetes	0.2–2 µm × filamentous	1,000,000
Fungi	5–15 µm × filamentous	100,000
Algae	10–40 µm	10,000
Protozoa	20–200 µm	1000
Nematodes	200–1000 µm	100
Other invertebrates	>200 µm	1

† For comparison, a square yard of soil 6 inches deep may contain 30 to 300 earthworms.
‡ Remember, there are 454 grams in a pound (a pound of soil is the approximate amount contained in a pint jar).

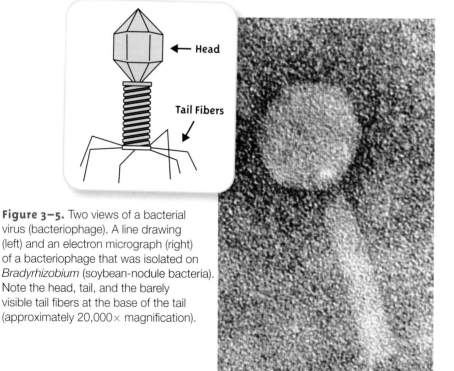

Figure 3–5. Two views of a bacterial virus (bacteriophage). A line drawing (left) and an electron micrograph (right) of a bacteriophage that was isolated on *Bradyrhizobium* (soybean-nodule bacteria). Note the head, tail, and the barely visible tail fibers at the base of the tail (approximately 20,000× magnification).

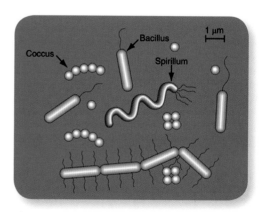

Figure 3–6. Common shapes of soil bacteria and archaea.

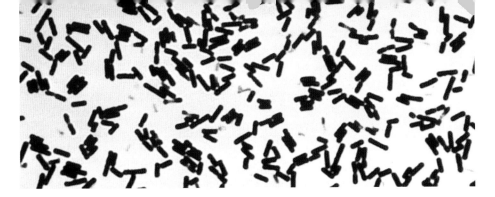

Figure 3-7. A view of *Clostridium perfringens*. These short, rod-shaped, spore-forming bacteria can be isolated from soil and can cause food poisoning in humans (1000× magnification).

similar size and shape, and do not have a cell nucleus or other membrane-bound organelles within their cells. Archaea have been found more recently, however, to have an independent evolutionary history with different biochemistry pathways, different cell wall and cell membrane compositions, and different genetics that separate them from the bacteria. Many of the archaea are difficult to culture in the laboratory, and this likely explains why little was known about them until the last several decades when DNA sequencing became available. Archaea are thought mainly to live in extreme environments, such as hot springs or saline pools, but also can be found in soil.

Classification of bacteria and archaea is complex and based on size, shape, and biochemistry (phenotypic classification) or genetics (phylogenetic classification). Size and shape variations range from round to short rods to spirals. Short rods dominate most soil populations. Some soil bacteria are pleomorphic in that they can change size and shape, usually as a function of age. With a light microscope, it is difficult to see anything other than the basic shape of the cells (figure 3-7).

Individual bacteria and archaea can use a wide variety of substrates as an energy source (electron donor): organic or inorganic compounds, or even sunlight. As discussed above, just as all living organisms must have an electron donor, they also must have an electron acceptor. For aerobic organisms, the electron acceptor is oxygen. Some bacteria and archaea, however, have developed metabolic systems where they can live in the complete absence of oxygen (anaerobic) using compounds such as nitrate, ferric iron, or sulfate as an electron acceptor. Organisms using oxygen gain more energy than organisms using other electron acceptors so, if present and the organisms can use it, oxygen offers an advantage. But in a waterlogged soil, oxygen is scare because oxygen movement through a pore filled with water is only about 1/10,000th the rate of oxygen movement through a pore filled with air, so anaerobic organisms will dominate.

CYANOBACTERIA are an interesting group of soil bacteria that get their energy in a unique way. They use sunlight and can photosynthesize like higher plants. Thus, unlike most bacteria and archeae, they live independent of soil energy sources. They

---

What Do All Organisms (People Included) Need for Life?

WATER. Water is an essential component of all living cells and often makes up more than 70% of an organism's weight.

ELECTRON DONOR (energy source). During respiration, electrons are transferred during metabolism, releasing energy for the organism. Plants in the presence of light get their electrons from water and release oxygen. For most consuming organisms, organic compounds containing reduced carbon are the source of electrons, but some microorganisms can get electrons from inorganic compounds such as ammonium, ferrous iron, reduced sulfur, and reduced manganese.

ELECTRON ACCEPTOR. For most organisms, the electron acceptor is oxygen (aerobic organisms), but some microorganisms (anaerobic organisms) can use other compounds such as nitrate, nitrite, ferric iron, and oxidized sulfur.

CARBON SOURCE. For most organisms, the carbon source is organic compounds, but plants and some microbes can use inorganic carbon dioxide or carbonates as a carbon source.

ELEMENTS. All organisms need to acquire elements from their environment such as nitrogen, phosphorus, potassium, calcium, magnesium, sulfur, iron, cobalt, copper, manganese, and molybdenum. Some elements, such as nitrogen and phosphorus, are needed in larger quantities and used to make proteins and nucleic acids; others such as iron and cobalt are needed in smaller quantities and required for enzymes to properly function.

GROWTH FACTORS. Some organisms—including ourselves—must have specific components such as essential amino acids, vitamins, nucleic acids, and essential fatty acids in their diet. Humans cannot make these materials, which are required for growth and development. Many microorganisms also need specific compounds in their diet, but others can manufacture what they need from simple inorganic compounds.

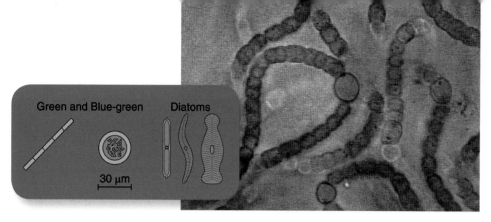

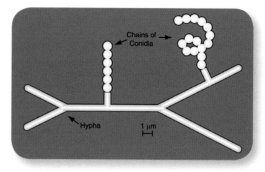

Figure 3–8. Examples of common phototrophic organisms in soil that use sunlight as an energy source. The diagram shows the shapes of cyanobacteria and the plant-like protists (algae). On the right above is a photomicrograph of the cyanobacteria *Anabaena*. Notice the greenish color due to the presence of chlorophyll. Also note the occasional larger round cells. These are heterocysts where nitrogen fixation occurs (1000× magnification).

Figure 3–9. Actinomycetes are sometimes called "thread bacteria" and grow in filaments or strands. A single strand is called a hypha; multiple strands are hyphae. Most actinomycetes reproduce by spores (*conidia*). This group of organisms gives soil its characteristic odor.

were formerly known as blue-green algae because they produce oxygen during photosynthesis much like plant-like protists (algae), but as prokaryotic cells, they are now recognized as bacteria. Several cyanobacteria are also able to fix atmospheric nitrogen (discussed later in this chapter) in specialized structures called *heterocysts* (figure 3–8). Members such as *Tolypothrix*, *Anabaena*, and *Nostoc* have been reported to fix 224 kilograms of nitrogen or more per hectare (200 or more pounds per acre) per year, especially in rice culture. Nitrogen fixation under row crops such as corn or cotton is much less, perhaps 11 kilograms per hectare (10 pounds per acre) per year.

ACTINOMYCETES are technically classified as bacteria but are such an important group in soil that many soil biologists discuss them separately. They are considered a transitional group between bacteria and fungi (figure 3–9 and 3–10). They possess characteristics similar to fungi in that they grow into filaments but are classified with bacteria because they have a prokaryotic cell structure (cells that do not have a nucleus membrane among other features). Numbers of actinomycetes that can be cultured from soil are often less than other bacteria but greater than fungi (table 3–1). In soils that are drier, have low levels of organic matter, or high pH, the numbers of actinomycetes may be greater than the numbers of other bacteria.

Only three genera of actinomycetes are commonly found in soil, and the dominant genus, *Streptomyces*, often accounts for 70% to 95% of all actinomycetal cells isolated from the soil.

The actinomycetes as a group grow relatively slowly and presumably do not grow as well as other soil bacteria and fungi when fresh organic materials are added. They do have the ability, however, to utilize very simple as well as highly complex organic molecules as a source of carbon and energy, and they play a significant role in decomposing more resistant organic compounds.

Soil actinomycetes, especially members of the genus *Streptomyces*, produce antibiotics. Many of these antibiotics, such as streptomycin, neomycin, and the tetracyclines, are important in treating human infections. Antibiotic production in ecological niches within the soil, however, is not well documented or understood, mainly because this happens within small microsites that are difficult to study.

Prokaryotic and Eukaryotic Cells

There are two basic cell types in the biological world: *prokaryotic* and *eukaryotic*. You are composed of eukaryotic cells; bacteria and archaea are prokaryotic. Prokaryotic cells are smaller and less complex. They have no membranes surrounding their nucleus or other internal organelles. Among other features, eukaryotic cells have more complex chromosomes (humans have 23 chromosome pairs), whereas prokaryotic cells have simpler chromosomal DNA (often a single chromosome composed of double-stranded DNA in a loop). Many scientists think eukaryotic cells evolved from prokaryotic cells.

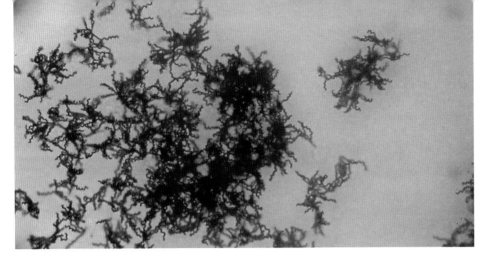

Figure 3–10. The aerial conidia of some Streptomyces form chains that may coil. Individual conidia are too small to be seen at this magnification but careful observation will confirm the spiral nature of the coils. These coils are sometimes called pigtails of conidia (500× magnification).

Presumably, an organism that produces an antibiotic that suppresses its neighbor has a competitive advantage in using the resources in its vicinity.

FUNGI are a huge group of eukaryotic organisms that range from single-cell microscopic yeasts to large multicellular mushrooms. Mushrooms (toadstools) may come to mind when you think of fungi, but these represent only a small portion of the kingdom (and the mushroom is only the small, above-ground fruiting structure of a larger soil organism that lives below ground).

As eukaryotic cells, fungi are more complex than prokaryotic bacteria and archaea. Many soil fungi develop a vast array of spreading filamentous structures, singularly called *hypha* and collectively called *hyphae*. These fine hair-like structures serve a common function as plant roots. The hyphae allow the fungus to spread rapidly and penetrate the soil, leaf litter, wood, or other substrates. Many fungi also have very complex means of reproduction, often by spores that much like plant seeds are a reproductive or resting stage. Spores often are better able to survive heat and desiccation than vegetative cells. The spores may be asexual or sexual (some fungi produce both). In asexual reproduction, a single hyphal strand generates the spores, often by budding or fragmentation. In sexual reproduction, two mating types join to produce the spores. Spores are dispersed by a number of ways, including water, wind, and animal movement; some spores even have flagella (whip-like structures) and can swim to new locations.

Some of the fungi commonly isolated from soil include *Aspergillus*, *Cladosporium*, *Fusarium*, *Penicillium* (figure 3–11), and *Trichoderma*. All fungi lack chlorophyll and cannot use carbon dioxide as a carbon source (that is, they must have organic carbon in their diet). Fungi respond rapidly when substrates are added to soil and appear to be dominant in the breakdown of cellulose, hemicelluloses, and lignin.

Fungi usually are not as numerous as bacteria and archaea in soil, but because they are larger, they often have more biomass (living tissue). On laboratory media a colony may grow from a vegetative cell or from a spore. The spore may germinate when placed in a suitable environment, but it may not be actively growing in the soil. Because fungi can grow from both vegetative cells and spores, the reproductive units are commonly called *propagules*. Soils typically have 10^4 to 10^5 fungal propagules per gram of soil.

As a group, fungi require oxygen (*aerobic*) as a terminal electron acceptor and tend to dominate in acid soils. This does not necessarily mean that fungi prefer acid soils: they simply can tolerate acidity better

Geosmin, the Smell of Soil

When you turn a spade of garden soil or step into woods after a rain, you notice a distinctive earthy odor. That smell comes from *geosmin*, a volatile organic compound produced by several soil organisms, most notably the actinomycetes. A culture plate of actinomycetes can smell just like freshly turned soil. So, why does outside air smell the way it does after a rain? The water enters the soil and displaces the soil air that contains the geosmin into the atmosphere. The human nose can detect geosmin at concentrations as low as 5 parts per trillion. It is unknown why soil actinomycetes produce this compound.

Streptomycin

Streptomycin is an antibiotic produced by soil actinomycetes of the genus *Streptomyces*. It was discovered by Selman A. Waksman and Albert Schatz in 1943, shortly after penicillin was discovered. Streptomycin was the first antibiotic effective in treating tuberculosis. It acts by inhibiting synthesis of vital cell proteins in the tuberculosis-causing bacteria.

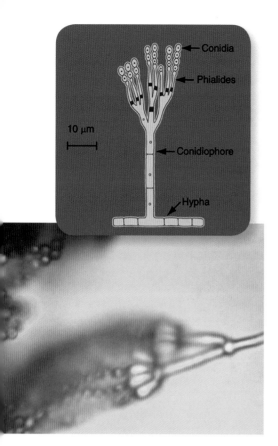

Figure 3–11. Some structural features of the fungus *Penicillium*. The spores (conidia) are like seeds and can give rise to new organisms. The lower image is a photomicrograph of the organism showing the actual features (1000× magnification).

than bacteria, archaea, actinomycetes, or cyanobacteria, and without the competition from other organisms in acid soil, the fungi have access to more resources, resulting in higher populations. Because fungi have an extensive hyphal network in soil, they are thought to be important in soil structure (the grouping together of soil particles). Also, some fungi can produce antibiotics: penicillin is produced by members of the genus *Penicillium*.

PROTISTS is a term used to described a large group of microscopic, eukaryotic, mainly unicellular, organisms found in soil as well as oceans and fresh water. The term comes from the Greek word *protos*, meaning first (Earth's first eukaryotes). Scientists use the term to describe simple, diverse, eukaryotic organisms that are not fungi, plants, or animals. Protists are often labeled according to their ecological roles, which include three major types: plant-like protists (algae), animal-like protists (protozoa), and fungus-like protists.

Soil *algae* are plant-like protists widely distributed in surface soils of the world. They contain *chlorophyll* (get their energy from light), generate oxygen, and get their carbon from carbon dioxide in the atmosphere. Algal numbers are usually greater in surface soils because of the abundance of light. They range from rather small, single-unit cellular forms, to larger filamentous forms (figure 3–8), and they are classified according to the types of pigments they produce, their storage products, and their cell size and shape. In temperate soils, the green algae (Chlorophyta) are the most common. Their metabolism is much like that of plants but on a smaller scale, and they are considered to be evolutionary forerunners of plants. Their numbers can be so high, especially under wet conditions with abundant phosphorus and other elements, that the surface of the soil turns shades of green.

Diatoms are a unique group of photosynthesizing protists that have distinctive cell walls consisting of overlapping halves impregnated with silica. The cell walls typically have bilateral or radial symmetry (figure 3–8) and are often beautifully ornamented. Although commonly isolated from soils, diatoms are known to be major producers in aquatic environments, serving as a base in food chains. Deposits of fossilized diatoms are mined as diatomaceous earth and used in abrasives and for filtering. Scientists have estimated that the plant-like protists (mainly aquatic diatoms and other algae) generate at least one-half of the oxygen in Earth's atmosphere.

Protozoa are animal-like protists common in soil; some consider them to be the beginning of the animal kingdom. Even though protozoa are larger than bacteria and archaea, ranging in size from 20 to 200 micrometers, one must still use a microscope to see them. Soil protozoa consume either organic matter or other microorganisms, mainly bacteria. In fact, some protozoa have a definite preference for individual species of bacteria, and it is estimated they must consume up to 10,000 bacteria to acquire the needed energy for one reproduction. Protozoa reproduce by splitting—the mother cell splits into two daughter cells.

Because the ability to move is so important to protozoa for obtaining food, they are classified based on how they move. The Mastigophora (figure 3–12) are flagellated and swim in water films by means of flagella (small whip-like structures). The Sarcodina have pseudopodia (false feet) and move by extruding protoplasm in the direction of movement. The Ciliophora move by cilia, which are short hairs surrounding the organism that beat in unison for movement. Protozoa are commonly present at 10^2 to 10^3 per gram of soil. The protozoa depend on water films in the soil for access to food. When soil conditions

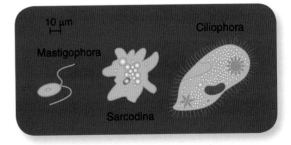

Figure 3–12. Three classes of animal-like protists (protozoa) in soil are broadly classified based on how they move.

are dry, several species have the ability to encyst themselves in a resting stage and wait until moisture returns.

The third group of protists consists of those that resemble fungi. These are less common in soil than plant- and animal-like protists. They use only organic materials as a carbon source (heterotrophic) and contain long, hyphal-like strands that look like fungal hyphae. Common members are *slime molds* that feed mainly on decaying organic matter or other microorganisms. As the name implies, during part of their life cycle they may form gelatinous-appearing slime. They can occur as a single-cell organism but when food is scarce, they congregate and start moving as a single body, often in brightly colored trails resembling fungal hyphae.

LARGER FAUNA

Soil animals play an active role in the biology of soil. They burrow through soil leaving pores, they help control the populations of other organisms, they produce organic compounds that bind soil aggregates together, and they feed on organic components of animal and plant residues to recycle nutrients. Some of the soil animals are permanent residents, whereas others only temporarily live part of their lives in soil. The former group includes nematodes, mites, springtails, insect larvae, earthworms, and larger animals.

The *nematodes* are the most numerous of the larger fauna (table 3–2), and their numbers far exceed other non-protozoan fauna. They are non-segmented worms (figure 3–13) that characteristically have spindle-shaped bodies. They swim through soil water films much like a snake swims through water, which is different from the movement of segmented worms (such as earthworms), which move more like inch worms. Common soil nematodes are 0.5 to 1.5 millimeters long, but some soil nematodes may be as long as 1 centimeter. Many can encyst themselves similar to the way protozoa do when the soil dries. Nematodes are noted for their parasitic attack on plants, but most in soil are beneficial.

Arthropods such as *mites*, *springtails*, and other larger organisms are invertebrate animals that have exoskeletons and jointed limbs. They are often involved in a soil food web, where the larger organisms feed on the smaller organisms (figure 3–14). A *Berlese funnel* (figure 3–15) is commonly used for extracting the more mobile fauna.

Worms are an extremely important group of soil fauna in soils that are not too acid. Some soil worms are too small to be readily seen with the unaided eye, such as *potworms*, but others, such as *earthworms*, are relatively large. Although total earthworm numbers are usually small, their large size per worm makes them a significant amount of the total faunal biomass.

Worms are active in decomposing dead and decaying plant remains. There are vertical burrowing and horizontal burrowing earthworms. Some earthworms, such as the night crawler (*Lumbricus terrestris*), come to the surface to obtain plant debris and drag the debris back into their burrows. They are important in the mixing action of soil and in soil aeration and water infiltration. You can find as many as 6,000 channels per acre (15,000,000 channels per hectare) coming to the surface in soils having high earthworm populations. Worms are a great source of food for larger animals such as birds and moles.

Table 3–2. Most soil fauna live in the top 10 cm of soil (expected populations are expressed in numbers per square meter).

Organism	Numbers per square meter†
Nematoda (nematodes)	5×10^6
Acarina (mites)	1×10^5
Collembola (springtails)	5×10^4
Enchytraeid worms (potworms)	2×10^3
Mollusca (snails and slugs)	1×10^3
Earthworms	1×10^2
Larger myriapods (millipedes and centipedes)	1×10^2

† Other organisms such as isopods (wood lice), ants, beetles and larvae, and spiders are expected at approximately 1000 per square meter.

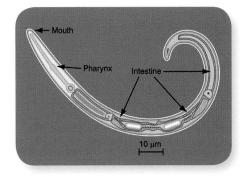

Figure 3–13. Nematodes are numerous in soil. Many consume smaller organisms and organic debris, but a few are plant parasites. Nematodes are sometimes called roundworms or eelworms because of their body appearance.

LIFE ABOVE THE SOIL DEPENDS ON LIFE IN THE SOIL

As we have seen above, soil contains a vast assortment of organisms that are active in the decomposition and recycling of elements. At any one time, significant quantities of the nutritive elements on Earth are tied up in living tissue. Without recycling, this finite quantity of elements would quickly become limiting. Soil is the natural medium where much of the decomposition by consumers occurs and where new growth by producers takes

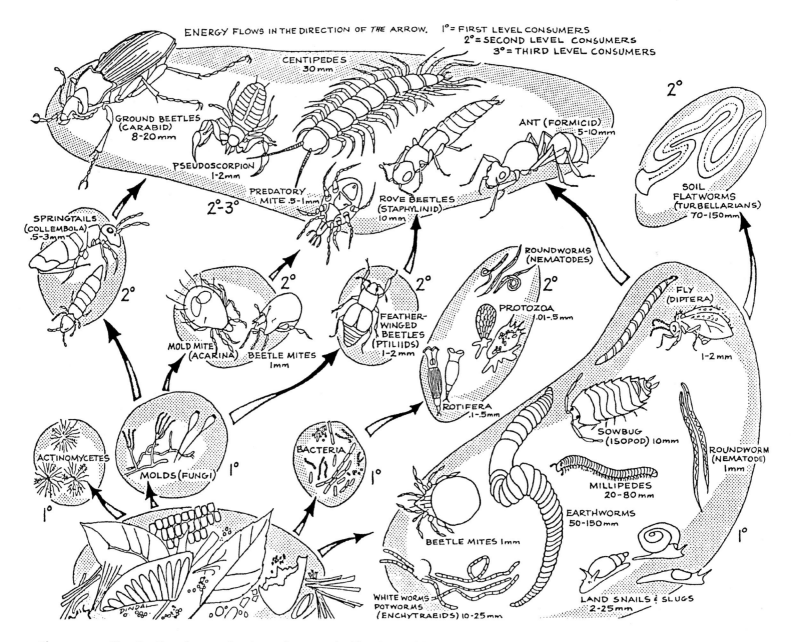

Figure 3-14. The direction of energy flow in a soil community. The plant residues in the lower left corner provide reduced carbon from the photosynthetic process. First-level consumers break down these residues. At the next level (second-level consumers), larger organisms consume the carbon contained in the smaller organisms, and so on up to the third level of consumers. The arrow to the right from the residues indicates that some larger organisms consume the plant residues directly and in the process break big pieces into smaller pieces to accelerate the rate of decomposition.

BERLESE APPARATUS

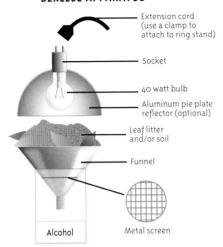

Figure 3–15. A Berlese Funnel is a common apparatus used for extracting soil fauna, especially those that readily move in the soil. A light bulb is positioned above a soil sample so that the surface of the soil is about 45°C. The heat, light, and drying conditions cause the organisms to move downward where they fall through the screen and collect in the alcohol (water can be used to view living organisms).

PLANT NUTRITION AND GROWTH

Soil provides anchorage, water, and nutritive elements to growing plants. A portion of the nutrients comes from the mineral weathering discussed in chapter 2, but others are released as plant and animal residues are decomposed. This occurs naturally in all soils of the world, and between weathering and nutrient recycling, most of a plant's nutrient needs in natural ecosystems are met. The exceptions, especially in agricultural soil, are the three nutrients often found in industrial fertilizers: nitrogen, phosphorus, and potassium.

Interestingly, microbial systems have evolved in nature to provide both nitrogen and phosphorus to growing plants. These beneficial systems involving biological partners presumably are an evolutionary response to these common plant deficiencies. These partnerships have evolved over eons of time. For example, fossil records indicate microbial partnerships with plants existed some 400 million years ago. (The third plant nutrient in fertilizers, potassium, is released during organic matter decomposition and mineral weathering, but doesn't depend on specific microorganisms to enhance its uptake by plants.)

As the human population is projected to increase from the current 7 billion (2012) to more than 9 billion by 2050, food production will need to increase accordingly, and these natural biological systems must be better understood and utilized to feed these additional people. We will briefly discuss the biological partners involved in plant use of nitrogen and phosphorus, and learn more about the chemistry of these and other plant nutrients in chapter 4.

NITROGEN is a major component of plant protein and enzymes and is required by plants in large amounts. Nitrogen is the most deficient nutrient in the world's soils, and more nitrogen fertilizers are applied annually than any other nutrient. The cereal grains (such as corn, wheat, rice, barley, and oats) serve as major food sources for many of the world's inhabitants, and these grains are particularly responsive to nitrogen fertilizers. Excessive application of fertilizers can result in nitrogen in water supplies, where it causes human and environmental problems. In addition, humans currently use large quantities of natural gas to make nitrogen fertilizers, and some ask whether this use is sustainable into the future.

Nitrogen: From Gas to Ammonia to Plants

Earth's atmosphere is 78% (by volume) nitrogen gas (N_2), but plants cannot use this form of nitrogen. It is a colorless, odorless, inert gas that consists of two nitrogen atoms bound together by a very strong triple bond. This triple bond must be broken—either by industrial processes or by biological processes in the soil—before plants can use the nitrogen.

When nitrogen fertilizers are made (Haber-Bosch process), the bonds are broken and the nitrogen is combined with hydrogen to form ammonia (NH_3). The hydrogen comes mainly from natural gas. It takes approximately 30 million BTUs of natural gas to manufacture one ton of nitrogen fertilizer. Worldwide, about 5% of the natural gas consumption, or somewhat less than 2% of the world's total energy consumption, is used to make nitrogen fertilizers.

Before nitrogen fertilizers were manufactured, microorganisms were the main source of nitrogen for crop plants. The microbial process is called biological nitrogen fixation: certain microorganisms take N_2 gas from the atmosphere and convert it into ammonia.

$$N_2 + 8H^+ + 8e^- \leftrightarrow 2NH_3 + H_2$$

The ammonia is then converted into organic forms plants can use.

The words ammonia (NH_3) and ammonium (NH_4^+) can be confusing. Ammonia is a gas, but it readily combines with water to form ammonium, which is a common form of nitrogen taken up and used by plants.

$$NH_3 + water \leftrightarrow NH_4^+ + OH^-$$

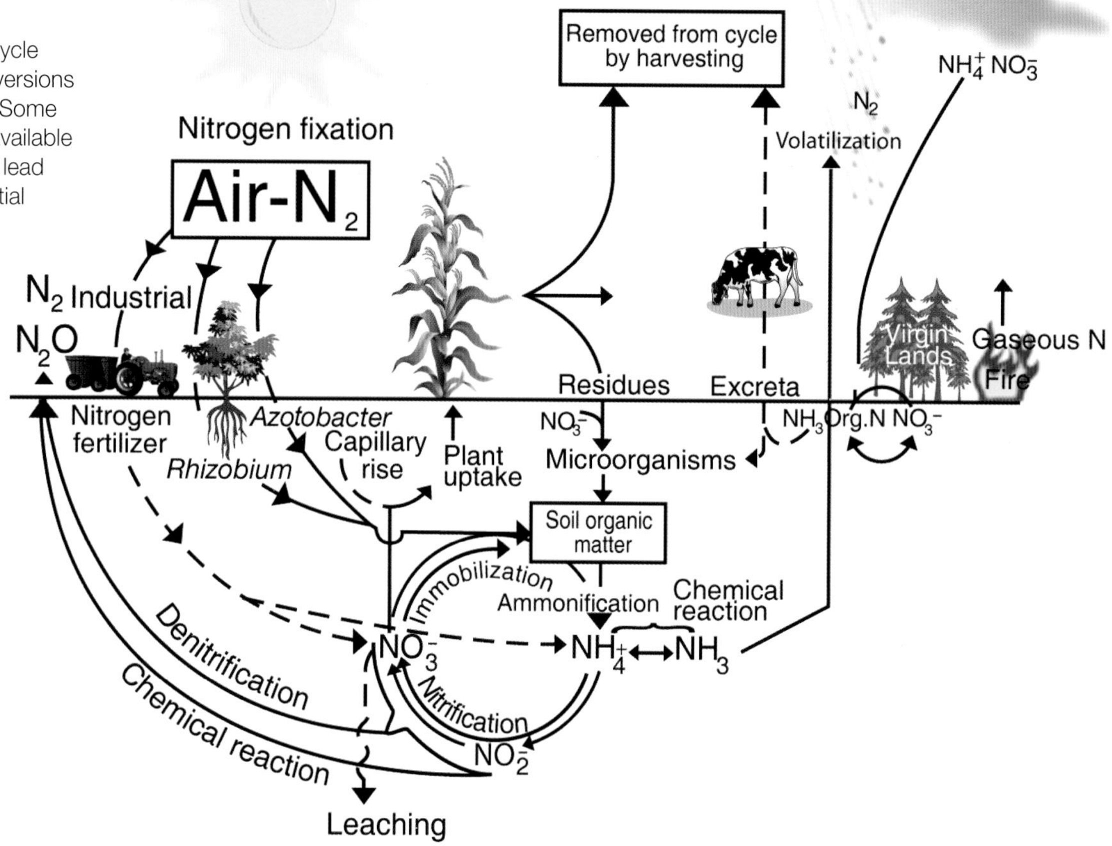

Figure 3–16. The nitrogen cycle involves many microbial conversions of various forms of nitrogen. Some conversions make nitrogen available for plant growth while others lead to nitrogen losses and potential environmental problems.

Soil microorganisms are important in the *nitrogen cycle*; they convert nitrogen into many forms, some of which are available to plants (figure 3–16). The nitrogen fixers are either 1) free-living or 2) live with higher plants in a *symbiotic association* where both partners benefit. The free-living organisms (example *Azotobacter* in figure 3–16) are usually responsible for relatively small amounts of fixed nitrogen in most soils (tens of pounds of nitrogen per acre per year). Microorganisms associated with higher plants (example *Rhizobium* in figure 3–16) can fix relatively large amounts (up to hundreds of pounds of nitrogen per acre per year).

There are two distinct types of symbiotic nitrogen-fixing associations in higher plants: bacteria (*Rhizobium* or *Bradyrhizobium*) that associate with legumes (such as soybeans, green beans, peas, clovers, peanuts, and alfalfa) and bacteria (*Frankia*) that associate with shrubs (including alder, silverberries, and mountain mahoganies). In both associations, nitrogen gas from the atmosphere is reduced to ammonia within the plant by the enzyme *nitrogenase*, which is produced by the bacteria.

It is truly a symbiotic association (*mutualism*) because the plant benefits from the nitrogen and the microorganism benefits from the sugars provided by the plant. The bacteria live within special root structures called nodules (figure 3–17). Scientists have extensively studied *biological nitrogen fixation*. Only prokaryotic cells (bacteria and archaea) can fix nitrogen from the atmosphere, although researchers are trying to genetically engineer plants, such as corn and wheat, to fix their own nitrogen independent of microorganisms. If successful, the need for expensive nitrogen fertilizers may be reduced or eliminated.

Once plants and soil organisms have died, the nitrogen they have assimilated in organic forms is released into the soil during decomposition. The conversion of organic forms of a nutrient to inorganic forms is called *mineralization*. Once the ammonium is released, plants may use it, or if it remains in the soil, microorganisms during *nitrification* will convert the ammonium into the nitrate form (see the equation earlier in this chapter). Both ammonium and nitrate are useable by plants, but as an anion, the nitrate form is much

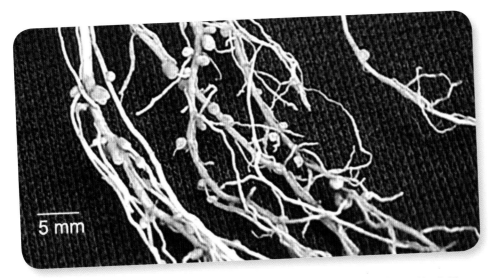

Figure 3–17. Root nodules that harbor nitrogen-fixing beneficial bacteria (*Rhizobium*). The plant is a forage legume known as birdsfoot trefoil.

more subject to leaching losses (see chapter 4). Furthermore, the nitrate form is subject to *denitrification* losses when soils become waterlogged.

PHOSPHORUS is the second key element in plant nutrition with a microbial connection. Microorganisms help plants acquire phosphorus through several means. First, they help mineralize organic forms of phosphorus in similar fashion to that of nitrogen. Second, soil microorganisms also may be involved in releasing phosphorus from mineral forms by excreting organic acids that help dissolve phosphorus. Third, fungi and plants have developed a symbiotic relationship (mutualism) that aids in plant nutrition. The fungal–plant associations are called *mycorrhizae*, which literally means "fungus roots." Mycorrhizae allow for a trade where the fungus receives reduced carbon (sugars) from the plant, and the plant receives phosphorus (and other nutrients, water, and disease protection) from the fungus.

Phosphorus is not only deficient in many soils of the world, it is also relatively immobile, and roots must intercept phosphorus for it to be taken up by the plant. To explore a large volume of soil, beneficial mycorrhizal fungi can help. The extensive network of fungal hyphae radiating outward from the plant root goes beyond the root hairs of the plant and exploits a greater volume of soil, thus allowing greater uptake of nutrients, especially phosphorus.

There are two major types of mycorrhizal associations with plants (figure 3–18). The first is an *endomycorrhizal* association, in which the fungus grows inside the root cortical cells (plant cells beneath the surface cells of the root). The second is an *ectomycorrhizal* association, in which the fungus grows on the root surface and among root cortical cells.

Ectomycorrhizal fungi form associations with only about 5% of vascular plants, mostly on trees and shrubs (figure 3–19), although the host specificity varies among the fungi. *Ectomycorrhizae* are common in forests and one plant may form mycorrhizal associations with several fungi concurrently. As these fungi grow and colonize many trees within a forest, they may be among the largest organisms found on Earth.

Endomycorrhizal fungi commonly form vesicles within root cortical cells (figure 3–20). The *vesicles* are globular structures that serve predominantly as storage organelles. Also formed are *arbuscules* that look like small trees within the cell. These structures facilitate exchange of sugars and nutrients between the fungus and the plant. Arbuscular mycorrhizae are known to be associated with most world plants including *angiosperms*, *gymnosperms*, and ferns.

SOIL QUALITY AND SOIL MANAGEMENT

As people grow crops and manipulate soils, the soils tend to degrade with time, lowering their quality. *Soil quality* is a hard term to define. Perhaps the most commonly accepted definition is provided by the Soil Science Society of America: "The capacity of a soil to function within ecosystem boundaries to sustain biological productivity, maintain environmental quality, and promote plant and animal health." Soil quality involves physical, chemical, and biological properties.

Denitrification: From Plant-Available to Plant-Unavailable Nitrogen

Denitrification is a biological process where numerous soil organisms have the ability to use nitrate as a terminal electron acceptor, releasing dinitrogen (N_2) gas into the atmosphere. This only occurs in wet environments where oxygen is limiting (usually a flooded or waterlogged soil). Because this is a respiratory process, large losses of plant-available nitrogen can occur in a relatively short period. If the denitrification process is incomplete, a gas called nitrous oxide (N_2O) escapes into the atmosphere. This gas is a potent greenhouse gas, with 310 times the warming potential of carbon dioxide.

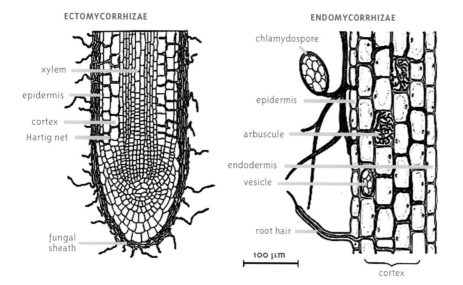

Figure 3–18. The two major types of mycorrhizal associations. The fungi of ectomycorrhizae (left) do not penetrate cortical cells while the fungi of endomycorrhizae (right) do.

Among some of the properties influenced by soil biological processes are soil organic matter, earthworms, soil structure, the amounts of carbon and nitrogen in soil microorganisms, the rate at which soils release plant-available nitrogen, soil respiration, and microbial diversity. Pesticides, where used, may negatively affect soil biology. Adding a variety of plant and animal residues, composts, mulches, or other organics to the soil is a major factor in favoring good biological activity of a soil.

SOIL ORGANIC MATTER

Microbes convert fresh plant materials and animal wastes into *soil humus*, the more stable portion of *soil organic matter* (figure 3–21). In natural ecosystems, fresh

Figure 3–19. Ectomycorrhizae on white pine show the typical multiple branching of root tips with the fungus covering the tips like a glove covers fingers on the hand. The hyphae that radiate outward into the soil were broken during plant removal from the soil (100× magnification).

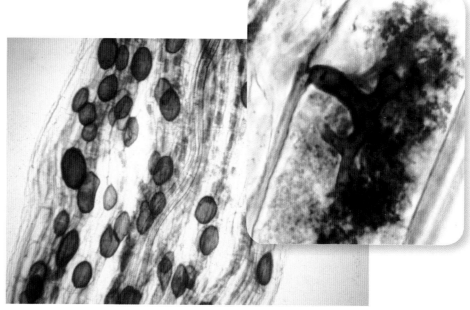

Figure 3–20. The endomycorrhizal association of soybean. The upper view shows vesicles (round structures), which are thought to mainly be storage organelles. Arbuscules can faintly be seen in the background (the ill-defined blue areas among the vesicles (100× magnification). The inset view shows an enlargement of an arbuscule where the exchange of sugars from the plant and phosphorus from the fungus occurs (1000× magnification).

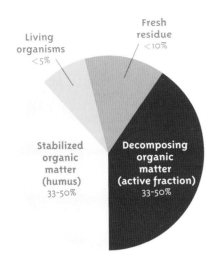

Figure 3–21. Organic matter is the main energy source for most soil organisms. The stable organic matter (humus) is relatively unavailable; the other pools are in various stages of being decomposed.

organic matter is constantly being added to the soil surface by plant leaves, for example, and to the subsoil by roots growing and dying. In agricultural systems, adding new organic residues each year is perhaps the best way to improve organic matter levels in the soil. Soil organic matter helps maintain soil structure, reduces erosion, improves soil nutrient and water-holding capacities, and serves as a food source for microorganisms. High crop yields and high productivity enhance the organic matter, since healthy, robust plants produce large amounts of roots and above-ground residues to be worked back into the soil. Adding compost or animal manures also is beneficial.

Conversely, tilling the soil, adding lime to acid soils, draining soils, and chopping up the organic residue and working it into the soil all tend to accelerate the rate of breakdown of soil organic matter. Thus, the level of organic matter in the soil at any one time is a balance of what is being added and what is being decomposed.

Converting to *no-till* or *reduced tillage* systems usually improves soil organic matter levels. Also, using rotations, especially those that include forage crops such as alfalfa or clovers, improves soil organic matter levels. Forage crops continue to grow early in the spring and late in the fall and involve no tillage. Keeping the soil covered with residue or growing plants reduces susceptibility to water and wind erosion, and improves soil organic matter levels by reducing losses.

EARTHWORMS

Earthworms are "nature's plow" and are active in turning, mixing, and aerating the soil (figure 3–22). Soil management practices that encourage earthworm populations include liming an acid soil, reducing tillage, reducing compaction, and adding manure and other plant residues that feed the worms. Surface mulches benefit worms by providing food and by retaining surface moisture. If carbon-to-nitrogen ratios of residues are high, adding non-acid-forming fertilizers such as calcium nitrate is beneficial because the worms need enough nitrogen to decompose organic compounds with high carbon content.

Soil management practices that harm earthworm populations include allowing

Organic Matter and Humus

Soil organic matter and soil humus are both the organic component of soil. Organic matter is all inclusive and includes living and dead animal and plant materials. Soil *humus* (pronounced "HEW mus", a Latin word meaning ground or earth) is the fraction of soil organic matter that has undergone extensive decomposition and represents the more stable fraction of soil organic matter or soil carbon.

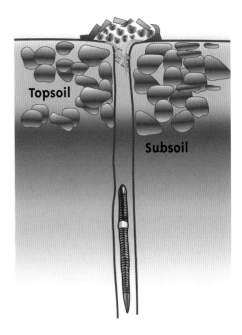

Figure 3–22. Earthworms are important in mixing materials into the soil. Some deep-burrowing worms such as nightcrawlers shown here drag surface organics into their burrows, which can extend 5 to 6 feet deep into the soil.

the soil to become too acid, allowing a soil to become poorly drained or waterlogged, burning crop residues, intensive tillage, and compacting the soil. Cropped fields usually have lower earthworm populations than pastures. Some herbicides are non-threating to worms, whereas other pesticides such as many fumigants can be quite harmful. Worms normally escape freezing and drying conditions by burrowing deeper into the soil.

SOIL STRUCTURE

A soil with high microorganism and soil fauna activities develops good soil structure. The microorganisms produce gums and gels (for example, extracellular polysaccharides) that help bind soil particles together, and some organisms, especially the fungi, produce extensive hyphae

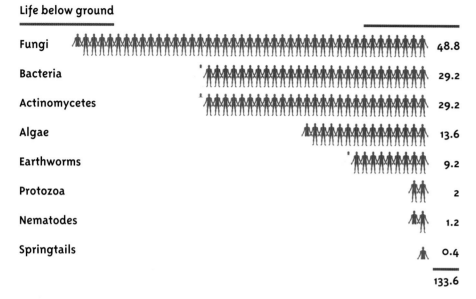

Figure 3–23. The mass of underground life to support life above ground. With a diversified diet (including meat), a typical North American needs approximately one acre of arable land for support. This figure illustrates the mass of a 200-pound person (mass above ground) versus the mass of organisms living below ground for that acre.

(filaments) that hold soil particles together. So, feeding the soil organisms by adding more plant and animal residues, composts, and mulches helps promote their growth and activity, often resulting in improved soil structure. This then helps the soil to function better physically (see chapter 2).

MICROBIAL BIOMASS

Microbial *biomass* is the living mass of organisms in the soil (figure 3–21). Having the soil in good physical and chemical condition and then feeding the microorganisms with organic residues is essential for keeping large and diverse populations of microbes alive. The microorganisms are responsive to added residues, and when conditions are favorable, biomass increases can reach their maximum within just a few days. Microbial biomass in soil can be quite extensive (figure 3–23).

MICROBIAL DIVERSITY

A diversity of soil organisms can decompose a variety of organic compound additions to soil and help control crop and soil pathogens. Crop rotations and mixed plantings provide different groups of organisms with different food sources and different environments in which to grow. This allows for ecological diversity.

As plant types change across the landscape and over time, the microorganisms associated with the plants vary and overall microbial diversity increases. An environment with a diverse microbial population is said to be more stable and responds more quickly to changes imposed on the landscape. The diversity includes larger organisms as well, and a soil with good diversity is truly a complex world (figure 3–24).

POTENTIALLY MINERALIZABLE NITROGEN

Nitrogen exists in the soil in many organic forms, as we learned above, some easy to decompose (or mineralize) and others that are quite resistant to decomposition. A soil that releases more nitrogen is more fertile, and the amounts released are quantified as potentially mineralizable nitrogen. As microorganisms decompose organic compounds, they release the contained nutrients. This is a measure of soil quality because nitrogen is so vital for plant growth. Soils with more organic matter and more microbial biomass will generally release more nitrogen than soils with low organic matter and little microbial biomass. But, because of the different components of soil organic matter (figure 3–21), one cannot just measure soil organic matter and assume a given fraction as potentially mineralizable nitrogen; it must be determined by a separate measurement.

SOIL RESPIRATION

Soil respiration is an overall measure of organismal activity in soil. With more life, there is more respiration, resulting in more carbon dioxide being liberated. Physical and chemical factors (such as temperature, moisture, nutrients, pH, or oxygen) may limit respiration, but the amounts of fresh, readily decomposable organic materials added to soil is often the deciding factor.

Plant roots directly respire carbon dioxide, but they also are important in providing a favorable environment for microorganisms. The zone surrounding a plant root (called the *rhizosphere*) is an energy-rich zone. The plant gives off exudates through its roots that microorganisms use. For example, in some situations plants can exude through their root system up to 25% of the energy flow from plant tops to roots (10% is common). Each

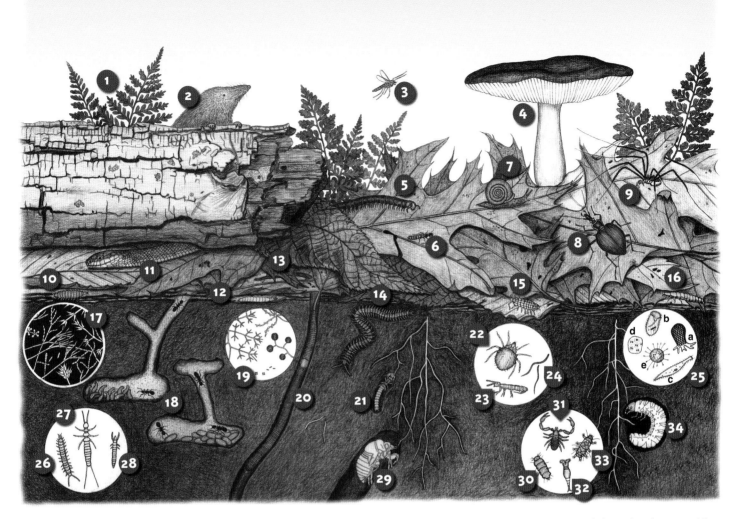

Figure 3–24. The complex world of living organisms in soil. Each organism lives in close proximity with others, and they often interact. All have their own place and function. Some are large and easily seen with the unaided eye, and others have to be magnified by 1000 times just to see their general shape. 1) Wood fern, 2) shrew, 3) march fly, 4) mycorrhizal mushroom (*Russula*) 5) millipede, 6) rove beetle, 7) snail, 8) snail-eating ground beetle, 9) daddy longlegs, 10) larva of soldier fly, 11) ringneck snake, 12) larva of crane fly, 13) soil flatworm, 14) soil centipede, 15) woodlouse, 16) larva of firefly, 17) filaments (hyphae) of soil fungi, 18) ants with their larvae and pupae, 19) bacteria and actinomycetes, 20) earthworm, 21) larva of click beetle, 22) oribatid mite, 23) springtail, 24) nematodes, 25) protozoa and algae: a, testate protozoa; b and c, ciliated protozoa; d, blue-green algae; e, heliozoa, 26) symphylan, 27) dipluran, 28) proturan, 29) cicada nymph, 30) pauropod, 31) pseudoscorpion, 32) rotifer, 33) tardigrade, 34) larva of scarab beetle.

plant contributes a unique root structure and type of residue to the soil and releases a wide range of organic compounds, including carbohydrates, amino acids, organic acids, enzymes, and vitamins. Because energy levels often limit growth of soil organisms, as discussed earlier, the exudates from plant roots are known to greatly stimulate biological activity and soil respiration.

PESTICIDES

In management of soils for plant growth, humans in relatively recent times have used *pesticides* to control weeds, insects, and disease. Some pesticides have chemical structures not naturally encountered in the environment, and some of these manufactured chemicals are slow to degrade in soil. Compounds may be slow to degrade because of their low solubility or their complex chemical structure.

Newer pesticides are formulated with their environmental impact in mind. Some earlier formulations were "hard" pesticides and had slow decomposition times in soil. An ideal pesticide should persist in the environment sufficiently long to affect its target organism but not so long as to harm non-target organisms. Decomposition by soil organisms is vitally important as a means of removing pesticides from the environment.

SUMMARY

The soil is a dynamic body continuously undergoing physical, chemical, and biological changes. Soil is alive with organisms, some small and some large, and is home for a huge complexity of life. The living organisms relate to conditions of the environment and contribute to the physical and chemical changes occurring within the soil. Microorganisms break down plant and animal residues, thus releasing and recycling nutrients. They also help break down pesticides. They aid plants in taking up nutrients, especially nitrogen and phosphorus. They are important contributors to the soil food web and in the process release carbon dioxide into the atmosphere. For most organisms, organic matter fuels the food web, and more organic materials added to the soil results in more soil life. Some organisms have evolved with unusual forms of metabolism where they can obtain energy from soil minerals through the oxidation of compounds such as nitrogen, sulfur, and iron. Many of the microscopic organisms of soil have yet to be grown in the laboratory, and their roles in soil await future discoveries in science.

Did you see the movie?

Soil organisms are living, breathing, moving creatures, and still photographs do not do them justice. To see movies, visit: http://www.agron.iastate.edu/~loynachan/mov/.

CREDITS

3–2, modified from AGI; 3–3, T. Loynachan; 3–4, James B. Nardi, University of Chicago Press; 3–5, T. Loynachan; 3–7, Centers for Disease Control and Prevention, image number 11196; 3–8 through 3–13, T. Loynachan; 3–14, Compost Science/Land Utilization, July/Aug. 1978, vol. 19, p. 9, JG Press, Emmaus, Pa); 3–16, modified from http://ohioline.osu.edu/aex-fact/0463.html; 3–17, T. Loynachan; 3–18, Arizona Cooperative Extension; 3–19, T. Loynachan; 3–20, T. Loynachan, D. Zuberer; 3–21, USDA; 3–24, James B. Nardi, University of Chicago Press. Chapter opener image, iStock.

Copyright © 2012. Soil Science Society of America, 5585 Guilford Rd., Madison, WI 53711-5801, USA. *Know Soil, Know Life*. David Lindbo, Deb A. Kozlowski, and Clay Robinson, Editors
doi:10.2136/2012.knowsoil.c3

GLOSSARY

ACTINOMYCETES Thread-like bacteria that give soil its "earthy" odor.

ALGAE Eukaryotic plant-like protists containing chlorophyll that can capture the energy of sunlight.

AMMONIA Form of nitrogen (NH_3) that converts in water to form plant-available ammonium (NH_4^+).

AMMONIFICATION The conversion of organic nitrogen to ammonium (NH_4^+) by the action of decomposers (bacteria).

AEROBIC In the presence of oxygen (O_2); requiring oxygen.

ANAEROBIC In the absence of molecular oxygen (O_2); not requiring oxygen.

ANGIOSPERM Flowering plant in which the mature seed is surrounded by the ovule (such as an apple, oak, daisy, grasses).

ANTIBIOTIC Organic compound that in low concentrations is inhibitory to other organisms.

ARBUSCULE Highly branched structure formed by endomycorrhizal fungi within root cortical cells.

ARCHAEA Prokaryotic single-celled microorganisms some of which live in extreme environments.

ARTHOPOD An invertebrate animal having an exoskeleton, a segmented body, and jointed appendages including insects, arachnids, and crustaceans.

ASSIMILATION Uptake or utilization of an element by an organism.

BACTERIUM (PLURAL: BACTERIA) Prokaryotic single-celled microorganism common in soil that lacks a nucleus or other membrane-bound internal structures.

BERLESE FUNNEL Method of extracting small animals from soil by use of a heat/light source.

BIOLOGICAL NITROGEN FIXATION Conversion by prokaryotic microorganisms of molecular dinitrogen (N_2) from the atmosphere to ammonia for organismal use.

BIOMASS Mass of living tissue.

CHLOROPHYLL Molecule that allows organisms, including plants, to use sunlight as an energy source.

CONIDIUM (PLURAL: CONIDIA) Resting structures of actinomycetes and fungi.

CYANOBACTERIA Bacteria that use sunlight as an energy source, may use N_2 from the atmosphere as a nitrogen source, and which evolve molecular oxygen (O_2) (formerly called blue-green algae).

DENITRIFICATION In the absence of molecular oxygen (O_2), some bacteria can use oxidized nitrogen (nitrate and nitrite) as a terminal electron acceptor and produce dinitrogen gas (N_2).

DIATOM Plant-like protists (algae) that have a siliceous cell wall and can use sunlight as an energy source.

EARTHWORM Segmented worm that burrows through soil leaving passages for rapid air and water movement.

ECOLOGY Study of how organisms relate to each other and to their environment.

ECTOMYCORRHIZA (PLURAL: ECTOMYCORRHIZAE) Association (usually beneficial) between fungi and plants that forms a fungal sheath on the outside of the root tip and hyphae among cortical cells of the root (adjective form is ectomycorrhizal).

ENDOMYCORRHIZA (PLURAL: ENDOMYCORRHIZAE) Association (usually beneficial) between fungi and plants that forms arbuscular and vesicular structures within cortical cells of the root (adjective form is endomycorrhizal).

EUKARYOTIC Organisms having more complex cell structure including membrane-bound internal organelles.

FAUNA Animal life.

FUNGUS (PLURAL: FUNGI) Large group of eukaryotic organisms that contain a rigid cell wall and include filamentous microorganisms, mushrooms, smuts, rusts, yeasts, and molds.

GEOSMIN Organic compound that is volatile and gives soil its "earthy" odor.

GYMNOSPERMS Plants whose seeds are not enclosed in an ovule (like a pine cone). Gymnosperm means "naked seed."

HETEROCYST Specialized cells of cyanobacteria where biological nitrogen fixation occurs.

HUMUS Dark, organic fraction of soil that has been well decomposed and is relatively stable.

HYPHA (PLURAL: HYPHAE) Long filaments or threads that are part of the structure of fungi and actinomycetes.

IMMOBILIZATION Conversion of plant available nutrients to an organic form which is unavailable to plants.

MACROORGANISM Organism larger than 2 millimeters.

MESOORGANISM Organism between 0.1 and 2 millimeters.

MICROORGANISM Organism smaller than 0.1 millimeters (100 micrometers).

MINERALIZATION Conversion of an element from an organic to an inorganic form.

MITE Small, eight-legged invertebrates, many of which are microscopic.

MYCORRHIZA (PLURAL: MYCORRHIZAE) Literally means "fungus roots." the association, usually beneficial, of specific fungi with the roots of higher plants (adjective form is mycorrhizal).

NEMATODE Nonsegmented worm, often microscopic. Some are parasitic on plants but many are normal soil inhabits that feed on fungi and bacteria.

NITRATE Plant-available form of nitrogen (NO_3^-).

NITRIFICATION Biological oxidation (loss of electrons) of ammonium to nitrite and nitrate.

NITROGENASE Active enzyme involved in biological nitrogen fixation that converts dinitrogen gas (N_2) into ammonia (NH_3).

NITROGEN CYCLE Describes the conversion of dinitrogen gas (N_2) from the atmosphere through microbes, plants, and animals to soil transformations and back to atmospheric dinitrogen gas. Important microbial processes include biological nitrogen fixation, mineralization, nitrification, and denitrification.

NO-TILL Practices by which a crop is planted directly into the soil with no primary or secondary tillage. The surface residue is left virtually undisturbed except where the seed is planted.

OXIDATION Chemically, the loss of electrons from an element.

PESTICIDE Compound, often organic, used to control or kill other organisms.

POTENTIAL MINERALIZABLE NITROGEN Organic component of soil nitrogen that is an active fraction easily mineralized or released by microorganisms.

POTWORM Microscopic, segmented worms living in soil that resemble small earthworms. They feed mainly on dead organic matter, bacteria, and fungi.

PROKARYOTIC Simple cell structure that has no internal organelles (including the nucleus) bound by membranes.

PROPAGULE Any cell unit capable of developing into a complete organism. For fungi, the unit may be a single spore, a cluster of spores, hyphae, or a hyphal fragment.

PROTISTS Diverse group of eukaryotic organisms that are neither animals, plants, nor fungi. They include animal-like (protozoa), plant-like (algae), and fungal-like (slime molds) organisms.

PROTOZOAN (PLURAL: PROTOZOA) Microscopic, unicellular, animal-like protists that are classified by their means of movement.

REDUCED TILLAGE A tillage system in which the total number of tillage operations preparatory for seed planting is reduced from that normally used on that particular field or soil.

REDUCTION Chemically, the gain of electrons by an element.

RESPIRATION Metabolism of an organism resulting in release of energy and carbon dioxide.

RHIZOBIA General term to identify root-nodule bacteria, such as rhizobium and bradyrhizobium, that are active in biological nitrogen fixation.

RHIZOSPHERE Zone of soil immediately adjacent to plant roots in which the kinds, numbers, or activities of microorganisms differ from that of the bulk soil.

SCIENTIFIC NOTATION Method used by scientists involving powers of 10 to work with very large or very small numbers.

SLIME MOLD Fungus-like protists that during part of their life cycle when food is short can move in a gelatinous trail or "slime" pathway.

SOIL QUALITY Generalized term used to describe a soil's ability to properly function and sustain plant, animal, and microbial productivity and maintain or enhance air and water quality.

SOIL ORGANIC MATTER Component of soil that contains carbon and is living or once was living but now is in various states of decomposition.

SPRINGTAIL Six-legged mesofauna or macrofauna that feed on organic materials or other smaller organisms. Most in soil are too small to be seen with the unaided eye.

SYMBIOTIC ASSOCIATION Relation between two different organisms where the activity of one affects the other. Three types exist:

i. Parasitism—one benefits, other is harmed (ticks)
ii. Mutualism—both benefit (i.e., Mycorrhizea)
iii. Commensalisms—one benefits other unaffected

VESICLE Round or vase-shaped organelle of endomycorrhizae used mainly for storage of lipids.

VIRUS Small particle that contains RNA or DNA and infects a living host cell.

CHAPTER 4

CHEMICAL PROPERTIES OF SOIL: SOIL FERTILITY AND NUTRIENT MANAGEMENT

CHAPTER AUTHORS

JOHN HAVLIN
BIANCA MOEBIUS-CLUNE

As we've seen, soil is the ultimate source of our food, fiber, and shelter, and soil's ability to produce these necessities depends on its fertility—the ability to provide nutrients to plants. The last chapter focused on the biological properties of soil; this chapter discusses its chemistry, particularly as it applies to agriculture.

All nutrients in our food originate from the soil. To stay healthy, humans need to acquire essential nutrients from many different food sources. Plants are a major direct human food source, but animals eaten by humans also obtain nutrients from a variety of plants (forages and grains) in their diets. Most of these nutrients come from agriculturally managed soils. But even foods caught from fresh and salt water contain nutrients that originate from soil—these are carried into water bodies as rainfall carries nutrients downstream to lakes and oceans. This transport is essential in appropriate quantities, but excessive transport of nutrients from soils into surface waters can have harmful effects on both agricultural and natural ecosystems (see chapter 6).

The demand for food and other products from agricultural systems will increase over the next decades. Any increase in agricultural production requires additional nutrient supply. Why? Because nutrients leave an agricultural field with the crop that is removed at harvest time. The natural supply of nutrients in the soil must then be amended with the amount removed: the exported nutrients need to be replaced with inorganic fertilizers and/or recycled organic materials to maintain soil fertility and nutrient supply to crops.

We need to manage nutrients well so that our slowly renewable soils can supply the needs of our growing population and other life on Earth. Therefore, we need to understand nutrient behavior in soils so that we can provide the nutrients that crops need and minimize the environmental risk of nutrient use.

The following sections include some of the techniques farmers and soil scientists use to improve nutrient availability and soil fertility. The management practices can enhance farm profitability and protect the environment, both of which are essential to secure the sustainable supply of safe, nutritious food and other agricultural products.

What Do Plants Need?

All organisms require water, carbon, and energy, as we have already seen in this book. Plants have several additional requirements for growth.

An element is considered essential to a plant if the plant cannot complete its life cycle without this element, and the deficiency can only be corrected by supplying the element. Seventeen essential plant *nutrients* have been established (see table). Carbon (C), hydrogen (H), and oxygen (O) are not considered "nutrients" per se, but are the most abundant elements in plants. Photosynthesis in green leaves converts carbon dioxide (CO_2) and water (H_2O) into simple carbohydrates (sugars), from which proteins and other organic compounds (the components of our foods) are synthesized. The supply of carbon dioxide and water rarely limits photosynthesis, although water stress is a frequent limit to optimum plant growth.

The remaining 14 essential elements are plant nutrients, classified as macronutrients and micronutrients. This classification is based on their relative abundance in plants. The *macronutrients* are nitrogen, phosphorus, potassium, sulfur, calcium, and magnesium. Compared with the macronutrients, the concentrations of the *micronutrients*—iron, zinc, manganese, copper, boron, chloride, molybdenum, and nickel—are very small. Micronutrients are often referred to as minor elements, but this does not mean that they are less important than macronutrients. Micronutrient deficiency or toxicity can reduce plant yield just as macronutrient deficiency or toxicity does.

RELATIVE PLANT NUTRIENT CONCENTRATIONS

	Name	Symbol	Form commonly used by plants
	Hydrogen	H	H_2O
	Carbon	C	CO_2
	Oxygen	O	O_2
Macronutrients	Nitrogen	N	NO_3^-, NH_4^+
	Potassium	K	K^+
	Phosphorus	P	$H_2PO_4^-$, HPO_4^{-2}
	Calcium	Ca	Ca^{+2}
	Magnesium	Mg	Mg^{+2}
	Sulfur	S	SO_4^{-2}
Micronutrients	Chloride	Cl	Cl^-
	Iron	Fe	Fe^{+2}, Fe^{+3}
	Boron	B	H_3BO_3
	Manganese	Mn	Mn^{+2}
	Zinc	Zn	Zn^{+2}
	Copper	Cu	Cu^+ Cu^{+2}
	Nickel	Ni	Ni^{+2}
	Molybdenum	Mo	MoO_4^{-2}

NUTRIENT SUPPLY TO PLANTS AND NUTRIENT CYCLING IN SOIL

All nutrients move or cycle through the environment, including in agricultural soils. Understanding nutrient cycling—how nutrients move through the soil, plants, and the atmosphere—is essential to efficient nutrient management. The specific biological, physical, and chemical processes each nutrient takes part in vary depending on the nutrient. We can visualize a general nutrient cycle in soil (figures 4–1 and 4–2) to illustrate some basic concepts. The next sections will then provide greater depth on some of the chemical processes involved.

As plant roots take in water, nutrients dissolved in the soil water enter the plant roots (figure 4–2, arrow 1). As plants absorb nutrients, complete their life cycle, and die, nutrients in the plant residues are returned to the soil (arrow 2). Dissolved in the soil water as ions, the nutrients react with the surfaces of soil mineral particles and soil organic matter, including plant residues.

When nutrients are removed from soil solution (plant uptake), several reactions occur to *buffer* or resupply nutrients to the soil solution. Ions *adsorbed* to the surface of clay minerals *desorb* from these surfaces to resupply the soil solution (arrow 4). This exchange (*adsorption* and *desorption*) in soil is a critical chemical reaction for making nutrients available to plants.

Nutrients added as minerals, organic materials, or other inputs increase the ion concentration in soil solution. Although some of the added ions remain in solution, others are *adsorbed* to mineral surfaces (arrow 3) or *precipitated* as solid minerals (arrow 5). Soil minerals can also *dissolve* to resupply the soil solution (arrow 6).

As soil microorganisms degrade plant residues, they can absorb nutrients from soil solution into their cells as needed (arrow 7). When plants, microbes, or other organisms die and decompose, the process of decomposition releases nutrients back to the soil solution (arrow 8). Biological processes such as these are important for plant nutrient availability, as well as for other properties related to soil productivity (discussed in chapter 3). Biological activity depends on numerous environmental conditions and on an adequate supply of organic carbon from plant residues or other added organic material. As they digest plant residues, soil organisms use oxygen and respire carbon dioxide (arrows 9 and 10).

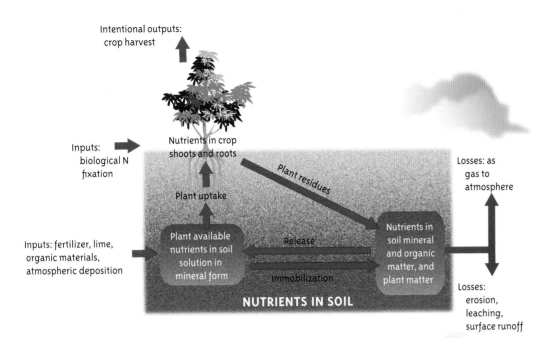

Figure 4–1. A generic nutrient cycle in an agricultural field. Nutrient inputs include the natural process of biological N fixation and the addition of agricultural inputs like fertilizer. Nutrients in soil can be taken up by plants if they are in a form the plant can use. Then they are removed by crop harvest, or returned to the soil organic matter pool through plant residues. Soil nutrients can be lost to the atmosphere as gas or lost through erosion, leaching, and runoff.

Numerous environmental factors as well as human activities influence nutrient concentrations in soil solution and affect mineral and biological processes in soils (arrows 11 and 12). For example, adding phosphorus fertilizer to soil initially increases the phosphate ($H_2PO_4^-$) concentration in soil solution. With time, $H_2PO_4^-$ decreases as plants take in the nutrient (arrow 1), as $H_2PO_4^-$ adsorbs on mineral surfaces (arrow 4), and as the phosphorus mineral precipitates (arrow 5).

All of these processes and reactions are important to plant nutrient availability; however, depending on the specific nutrient, some processes are more important than others. For example, microbial processes are more important than mineral surface exchange reactions for the avail-

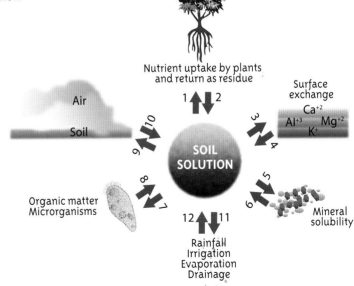

Figure 4–2. Various soil components and processes that influence plant nutrient concentration in the soil solution. Can any of these be influenced by management? Here's a hint: in figure 4–1 we saw that erosion can result in loss of nutrients.

ability of nitrogen and sulfur, whereas the opposite is true for potassium, phosphorus, calcium, and magnesium.

It is important to understand that an agricultural field is not a closed system. As crops are grown, harvested, and moved from the farm to the table (arrow 1), any nutrients taken up by these crops from the soil are also exported. If soil fertility is to be maintained for sustained crop growth, these exports must be balanced with imports of nutrients (arrows 2) in the form of added fertilizer and organic materials.

Other unintentional exports can occur that must also be balanced with imports (arrows 3 and 4, see figure 4–1). Losses of some nutrients occur in the form of gases that return to the atmosphere, and all nutrients (to varying extents) can be lost through erosion, surface runoff, or leaching to groundwater. Nutrients added beyond what plants need and beyond the soil's ability to retain nutrients ultimately enter surface or groundwater and degrade water quality. These losses are influenced by biological, chemical, and physical processes, depending on the nutrient and its form. Good soil management can help avoid or minimize some of these losses, and maximize the proportion of applied and soil-derived nutrients absorbed by the plant.

NUTRIENT MOBILITY IN SOIL

Now that we have a general picture of the nutrient cycles in soil, we consider the important question of how easily nutrients move through soils (nutrient mobility) in the soil solution, which strongly influences nutrient supply to plant roots.

Plants take up nutrients as ions in the soil water, as we learned above. The ions carry a positive charge (*cations*) or negative charge (*anions*) (table 4–1) and are absorbed from the soil solution by plant roots (figure 4–1 and 4–2).

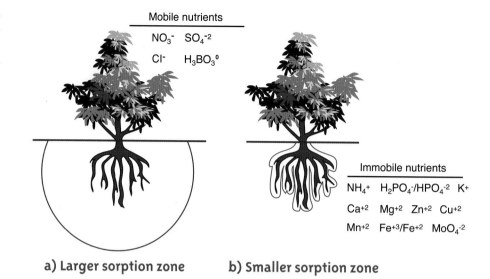

a) **Larger sorption zone**   b) **Smaller sorption zone**

Figure 4–3. a) Mobile nutrients can easily move to root surfaces from throughout the volume of soil occupied by roots. b) Immobile nutrients do not move easily, so roots can only access these nutrients if they are close to the root surface.

Mobile anions (NO_3^-, SO_4^{-2}, Cl^-) readily move through the root zone with water because they are not strongly attracted to *exchange sites* (soil particles) and are quite soluble in soil water. Since mobile nutrients are readily transported in soil water, they are available within the whole soil volume occupied by the plant root system (figure 4–3a). In contrast, immobile nutrients interacting with mineral and organic matter surfaces are less soluble, and do not readily move through the root zone to the root surfaces (figure 4–3b).

Of the nutrients classified as immobile, some are more mobile than others. Generally, ammonium (NH_4^+), potassium (K^+), calcium (Ca^{+2}), and magnesium (Mg^{+2}) are more soluble and mobile than the micronutrient cations, and much more mobile than phosphate ($H_2PO_4^-$) and molybdate (MoO_4^{-2}). Since these nutrients are

Table 4–1. Nutrients occur in the soil as cations and as anions.

Nutrient	Cations	Anions	Neutral
Calcium	Ca^{+2}		
Magnesium	Mg^{+2}		
Potassium	K^+		
Nitrogen (nitrate)		NO_3^-	
Nitrogen (ammonium)	NH_4^+		
Sulfur (sulfate)		SO_4^{-2}	
Phosphorus (phosphate)		$H_2PO_4^-$	
Zinc	Zn^{+2}		
Copper	Cu^{+2}		
Iron	Fe^{+3}		
Manganese	Mn^{+2}		
Molybdenum (molybdate)		MoO_4^{-2}	
Chloride		Cl^-	
Boron			$H_3BO_3^0$
Nickel	Ni^{+2}		

relatively immobile in soil, plant roots access these nutrients from a small volume of soil surrounding individual roots. As plants take up these nutrients, they create a small zone around their roots with very low concentrations of these immobile nutrients.

Understanding nutrient mobility in soils is essential to managing nutrient applications to maximize plant growth and recovery of applied nutrients. For example, nitrogen fertilizer can be broadcast across the soil surface or applied in bands with fairly similar results because nitrogen is mobile in soil. However, phosphorus is generally placed in concentrated bands in the root zone because it is relatively immobile in soil.

SOIL PROPERTIES INFLUENCING NUTRIENT SUPPLY TO PLANTS

As discussed in chapter 2, soils contain solids and pore space (air and water). The soil solids are sand, silt, and clay, as well as organic matter. The soil solids represent the source or natural supply of nutrients in soils and influence how nutrients behave when added to soil.

CATION AND ANION EXCHANGE

Solid particles have charged surfaces that attract cations and anions. The charged surfaces are important in retaining soil nutrients for plant use by a process called *ion exchange*. In soil, both negative and positive charges occur on the surfaces of soil particles and organic matter. Because of their large surface area and reactivity (see chapter 2), clay particles are more important in ion exchange than are sand and silt particles.

Cation (positive) and anion (negative) exchange occurs in soils between soil solution (water in the pores) and the charged surfaces of clay particles and organic matter. Ion exchange is a reversible process by which a cation or anion adsorbed on the surface is exchanged with another cation or

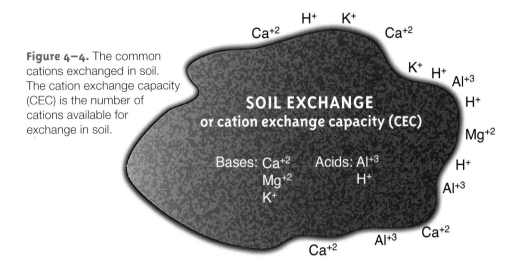

Figure 4-4. The common cations exchanged in soil. The cation exchange capacity (CEC) is the number of cations available for exchange in soil.

anion in the soil solution. Cation exchange is generally considered to be more important, since the *cation exchange capacity* (CEC) is much larger than the *anion exchange capacity* (AEC) of most agricultural soils. As soils form during the weathering process, new clay minerals create an excess negative surface charge; hence, in most soils cation exchange is much greater than anion exchange.

Most of the cations involved in ion exchange in soil are plant nutrients, except for hydrogen, aluminum, and sodium (H^+, Al^{+3}, and Na^+) (figure 4–4). Ion exchange reactions in soils are very important to plant nutrient availability and retention in soil. That's because the surface charge on soil particles allows the soil to store large quantities of nutrients and release small amounts into soil solution as these are depleted by plant uptake.

SOIL pH

As we just discussed, soil mineral and organic matter surfaces are charged and retain or exchange cations (mostly) and anions. Some cations are acidic (Al^{+3}, H^+, Fe^{+3}) while others are basic (Ca^{+2}, Mg^{+2}, Na^+, and K^+). The proportion of acidic and basic cations determines a soil's pH. In an acid soil the cation exchange sites are occupied with both basic and acidic cations. When the acidic cations increase relative to the basic cations, soil pH will decrease, or become more acidic. In soils that have a neutral (pH 7) or higher pH, the cation exchange sites are occupied only with basic cations (Na^+ may also be present).

The percentage of the exchange sites occupied by basic cations (Ca^{+2}, Mg^{+2}, and K^+) is the *base saturation* (% BS). The availability of Ca^{+2}, Mg^{+2}, and K^+ to plants increases with increasing base saturation. For example, a soil with 80% BS would provide these nutrients to growing plants far more easily than the same soil with 40% BS.

What Is pH Again?

The H^+ concentration or acidity of a solution is measured as pH, which is defined as follows:

$$pH = \log\frac{1}{[H^+]} = -\log[H^+]$$

Thus, pure water, which has a concentration of $H^+ = 10^{-7}$ M (M = molarity, moles/liter) has a pH of 7.0, which is a neutral (neither acidic nor alkaline) pH.

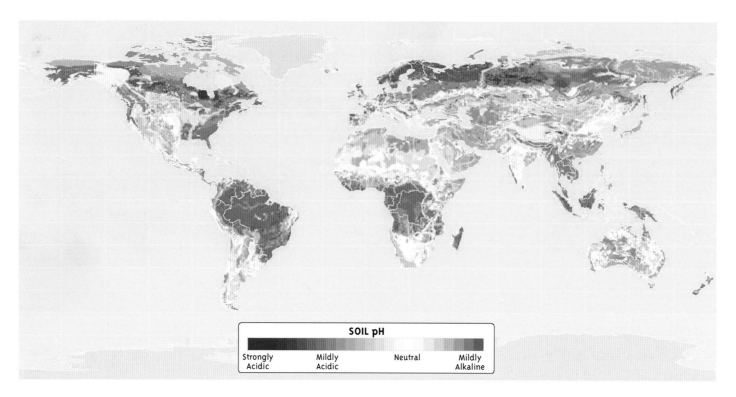

Figure 4–5. Acid and alkaline soil regions.

A soil with a high percentage of basic cations on the cation exchange surface will generally exhibit a higher soil pH compared with a soil with lower percentage of cations on the exchange surface (low % BS). Throughout the world, soil acidity (low pH) and alkalinity (high pH) problems influence plant growth. In the United States acid soils generally occur in regions where annual precipitation exceeds 63 centimeters (25 inches) per year, while alkaline soils tend to form in dry regions (figure 4–5).

Nutrients become less available to crops when soils are either too acidic or too alkaline. When soils are too acidic, potassium, calcium, and magnesium become less available, but aluminum can become too available, causing toxicity (figure 4–6). When soils are too alkaline, iron, manganese, zinc, and copper become less available. Phosphorus becomes less available with pH that is either too high or too low. Proper management of soil pH can help alleviate these soil problems. Farmers typically add lime (calcium-containing compounds) to fields to adjust for the acidification that normally occurs during crop production.

BIOLOGICAL PROCESSES: SOIL ORGANIC MATTER AND MICROBIAL ACTIVITY

Microbial activity and nutrient cycling through soil organic matter (see also chapter 3) substantially affect plant nutrient availability. Some microorganisms (such as *Rhizobia* and mycorrhizal fungi) directly aid plants in nutrient uptake through symbioses. Some plants exude organic acids that make certain nutrients more available. In addition, soil solution concentrations of nitrogen, sulfur, phosphorus, and several micronutrients are intimately related to microbial processes. These processes include organic matter decomposition and microbially controlled transformations in soils, such as denitrification (see chapter 3).

Soil organic matter comprises organic materials in all stages of decomposition. Fresh plant or other organic residues are

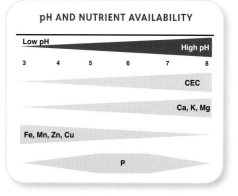

Figure 4–6. Nutrient availability at varying pH. When the band is thicker the nutrient is more available.

subject to fairly rapid decomposition. The climate, soil type, and soil and crop management all influence how much residue is produced and returned to the soil. In agricultural soils, crop residues ultimately decompose to form relatively more stable organic matter compounds. Such stabilized soil organic matter, also called humus, is the largest component of soil organic matter. It resists microbial degradation, and thus is able to perform essential roles such as maintaining optimum soil physical conditions important for plant growth, storing water (water-holding capacity), and storing and slowly releasing nutrients via the cation exchange capacity.

The primary soil microbial processes involved in fresh residue and organic matter turnover, or cycling, are *mineralization* (chapter 3) and *immobilization*:

$$\text{Organic complexed nutrients} \underset{\text{immobilization}}{\overset{\text{mineralization}}{\rightleftarrows}} \text{Inorganic nutrients in solution}$$

In the degradation process, organically complexed ions in the residue can be mineralized, or converted from organic to inorganic forms of the particular nutrient (i.e., nitrogen, sulfur, phosphorus). If there are insufficient nutrients in the residue to meet the microbial demand, then inorganic ions in the soil solution will be immobilized into the microbial cells. The microbial cycle of mineralization and immobilization occurs over a wide range of environmental conditions, but activity is maximized at about the same relative moisture and temperature conditions optimal for plant growth.

PHYSICAL PROCESSES: STRUCTURE, COMPACTION, AND WATER RELATIONS

Physical processes also drive nutrient availability, as they interact with biological and chemical processes in soil. Physical properties (chapter 2) influence both root access to nutrients, and the amount of environmental losses, because their effect on water storage and movement. Poor soil structure (loss of pore space) decreases the soil's ability to store and move water; good soil structure enhances it.

The problems associated with poor soil structure become worse in extreme weather. During drought, a poorly structured soil becomes harder and runs out of water sooner than a well-structured soil. In that case lack of water will prevent the plant from taking up nutrients. During heavy rainfall, since a poorly structured soil has fewer large pores, water is less able to infiltrate into the soil. More runoff will occur, which increases erosion, so nutrients and organic matter are lost from the surface where they are most concentrated. Not only are nutrients lost through this process, but some of the soil's ability to store nutrients via the cation exchange capacity is lost as well, so that future nutrient deficiencies are more likely.

Good soil structure provides adequate pores that allow roots to explore the soil volume, so that they have easy access to any available nutrients and water. Excessive tillage can cause poorly structured, compacted (hard) soil to form in the surface soil or the

How Does Rainfall Make Soils Acidic?

1. **With increasing rainfall amounts, more H⁺ is deposited in soil. How's that work?**

   Pure water undergoes slight dissociation:

   $$H_2O \rightleftarrows H^+ + OH^-$$

   The H^+ and OH^- concentrations in pure H_2O, when they are *not* in equilibrium with atmospheric CO_2, are 10^{-7} M, or a pH of 7.

   However, rain falls through the atmosphere and CO_2 dissolves in the water, so that the following reaction takes place:

   $$H_2O + CO_2 \rightleftarrows H^+ + HCO_3^-$$

   Water becomes acidic when in equilibrium with atmospheric CO_2. The pH of pure rainwater (without other pollutants) is ~5.7, so rainfall is a natural source of soil acidity that is important in the chemical weathering of parent materials. Burning fossil fuels adds sulfur (SO_2) and various nitrogen oxides (NO, NO_2, N_2O) to the atmosphere and turns rainwater even more acidic.

2. **With increasing rainfall, more water percolates through the soil profile, leaching and removing more of the basic cations.**

   The soil solution has to stay electrically neutral when water leaves the root zone as it percolates down to groundwater. The most soluble anions are NO_3^-, Cl^-, and HCO_3^-, while the most soluble cations are Na^+, K^+, Ca^{+2}, and Mg^{+2}. As anions leach, the basic cations also leach, reducing the base saturation, and leaving behind relatively more H^+ and decreasing the pH. Leaching occurs whenever there is more water than can be held in the soil pores of the root zone. Base saturation increases with increasing soil pH.

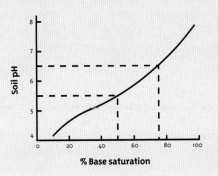

subsoil. Tilling when soil is too wet or with heavy equipment will compact soil. A compacted soil has lost its large pores. Roots then can't push aside soil solids, so they have difficulty growing through a compacted layer (figure 4–7). Soils can become so dense that roots are not able to branch out as they would in a well-structured and loose soil (figure 4–8). These roots are then prevented from exploring the full soil volume for nutrients and water. Plant nutrient deficiencies may result even when plenty of nutrients exist in the compacted soil.

ASSESSING NUTRIENT DEFICIENCY IN PLANTS

What happens to plants when soil nutrients are missing or in short supply? Plants tell us when they are deficient in one or more of their essential nutrients by displaying leaf colors or symptoms that are different from healthy plants.

Optimum plant health depends on the soil having an adequate supply of plant nutrients. We use *visual deficiency symptoms* (table 4–2) (and soil and plant analysis tools) to determine nutrient stress. Each nutrient *deficiency* exhibits a different symptom in a given plant, and different plants can show different symptoms with the same nutrient. Each visual symptom is related to nutrient function(s) in the plant.

Plants that are *severely* nutrient deficient exhibit a visual deficiency symptom. Plants that are *moderately* deficient usually exhibit no visual symptoms, although yield potential can be substantially reduced (figure 4–9). Correcting the deficiency by adding nutrients will maximize growth potential and increase plant nutrient concentration. *Luxury consumption* represents nutrient absorption beyond what is required for optimum growth, but is not detrimental to plant growth. In this case, as the supply of soil nutrients increases, the concentration of nutrients in the plant increases, but without an increase in plant growth.

Figure 4–7. Corn roots that were able to explore the full soil volume (left) and corn roots that could not grow deeper than about 20 cm because of a severely compacted subsoil layer (right). The roots on the left are able to access at least twice as much in terms of water and nutrients.

Figure 4–8. (left) This single root has struggled to grow into a poorly structured, compacted soil cannot access water and nutrients from most of the soil volume (left). (right) In a well-structured and loose soil, roots are able to proliferate. Dense exploration allows the roots to access water and nutrients in the entire soil volume easily.

Table 4–2. General description of nutrient deficiency symptoms in plants.

Plant nutrient	Visual symptoms
Nitrogen (N)	Light green/yellow leaves, especially older leaves; stunted growth; poor fruit development
Phosphorus (P)	Leaves may develop purple coloration; stunted plant growth and delay in plant development
Potassium (K)	Older leaves turn yellow initially around margins and die; irregular fruit development
Calcium (Ca)	Reduced growth/death of growing tips; poor fruit development and appearance
Magnesium (Mg)	Initial yellowing (older leaves) between veins spreading to young leaves; poor fruit development
Sulfur (S)	Yellowing of young leaves then whole plant; similar to N deficiency but occurs on new growth
Iron (Fe)	Initial distinct yellow/white areas between veins of young leaves
Manganese (Mn)	Interveinal yellowing or mottling of young leaves
Zinc (Zn)	Interveinal yellowing on young leaves; reduced leaf size
Boron (B)	Death of growing points, deformation of leaves with areas of discoloration

Nutrient deficiency symptoms appear when the soil nutrient supply is so low that the plant cannot function properly. In such cases, supplemental nutrients were needed long before the symptoms appeared. If the symptom is observed early, it might be corrected during the growing season with leaf or soil applications of fertilizer. However, yield is often reduced if adequate soil nutrients are not available when needed. Diagnosis of nutrient deficiency late in the growing season can still be useful in correcting the deficiency the following year.

More commonly, nutrient deficiency symptoms may not be visible, although nutrient levels may be considerably lower than required for optimum yield (*hidden hunger*). Soil and plant analyses are invaluable tools to identify hidden hunger, verify the specific nutrient causing the visual deficiency symptom, and guide farmers in managing soil nutrients to avoid yield loss from nutrient stress.

Identifying Nutrient Deficiency Symptoms Visually

Visual observation of the growing plant can help identify a specific nutrient stress. A nutrient-deficient plant exhibits characteristic symptoms because normal plant processes are inhibited. Visual nutrient deficiency symptoms are characterized by:

CHLOROSIS—uniform or interveinal yellowing or light green coloring of leaves

NECROSIS—death of leaf tips, margins, or interveinal regions of leaves

REDDENING—red or purple color on leaf margins, interveinal regions, or whole leaves as anthocyanins (red and purple pigments) accumulate

STUNTING—reduced plant height, shortened internodes; leaves remaining dark green or exhibiting light green or chlorotic symptoms

Unfortunately, if visual nutrient deficiency symptoms are apparent, a reduction in yield potential has already occurred. Therefore, quantifying the capacity of a soil to supply sufficient nutrients before planting or during the growing season and alleviating deficits before they decrease plant growth are essential for optimum plant growth and yield.

Zinc deficient

Nitrogen deficient

Phosphorus deficient

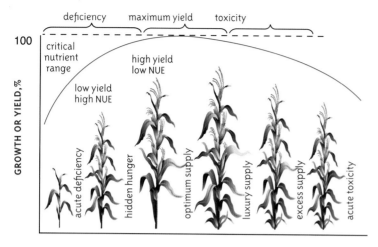

Figure 4–9. Optimum nutrient supply is essential for maximizing plant growth and nutrient use. Nutrient addition beyond a critical nutrient range does not increase plant growth and reduces nutrient use efficiency (NUE), which is the yield benefit gained from adding nutrients.

Soil Testing

Can you determine nutrient levels or availability to plants by measuring them in the soil?

Soil tests extract the 14 essential plant nutrients from soil samples to estimate nutrient availability to plants (carbon, hydrogen, and oxygen are not evaluated since they are abundantly available, as we learned earlier). Soil tests extract a part of a soil sample's total nutrient content that is related to (but not equal to) the quantity of plant available nutrients.

The objective of such testing is to predict the amount of each nutrient needed to supplement the native soil nutrient supply, so that plants have adequate amounts of all nutrients available during the growing season. For example, a soil testing "high" will require little or no additional supplement in contrast to soil with a low-test value. "Sufficiency levels" are commonly used in soil testing, where a high soil test represents 90% to 100% sufficiency in supplying adequate plant nutrients from the soil. Sufficiency levels decrease with decreasing soil test levels, and the probability of the plant responding to added fertilizer increases.

To collect soil for nutrient testing, a soil scientist uses a sampling tool to extract cores from a surface soil. Some 15 to 20 cores from each sampling unit get mixed in a bucket; then a subsample is removed and sent to a laboratory for analysis.

Identifying Nutrient Deficiency by Plant Analysis

Scientists perform plant analyses to verify the accuracy of an observed deficiency symptom and to identify plant nutrient shortages before they appear as symptoms.

Specific plant parts (for example, leaves or stems) are collected at specific times during the growing season. Samples are sent to a laboratory where the plant's nutrient content is determined. The specific laboratory performing the analysis usually provides sampling guidelines. To assist in diagnosis, plants from both deficient and normal areas are sampled for comparison.

A research technician measures leaf area of cotton before processing the leaves for analysis.

Figure 4–10. Broadcast or topdress application of nutrients (manure).

NUTRIENT MANAGEMENT

How do growers apply enough nutrients to ensure plants get what they need, while minimizing impact on the environment?

Proper nutrient management requires knowledge of the soil, the specific nutrient, the plant, and the environment. Once the right nutrient rate is determined (through soil testing), the grower must then consider how the nutrient is applied, especially for nutrients mobile in the soil (such as nitrogen). When nutrients are misapplied, there are costs for both the grower and the environment.

NUTRIENT APPLICATION TIMING

When to apply nutrients depends on the cropping system, climate, the specific nutrient, and soil. Sometimes nutrients are applied at times that may not be the most efficient for the plant, but are more favorably priced or better suited to the workload on the farm. Despite these considerations, growers should apply nutrients at a time that will maximize recovery by the crop and reduce potential losses to the environment.

NUTRIENT PLACEMENT

Proper placement of applied nutrients is as important as identifying the correct rate and timing. Placement decisions involve knowledge of crop and soil characteristics, since their interactions determine the nutrient's availability.

The options for fertilizing generally involve surface or subsurface applications before, at, or after planting. Placement practices depend on the crop and crop rotation, degree of deficiency or soil-test level, nutrient mobility in the soil, degree of acceptable soil disturbance, and equipment availability.

The simplest placement method is broadcasting, where nutrients are applied uniformly on the soil surface before planting, and can be incorporated by tilling or cultivating (figure 4–10). When the equipment is available, injecting nutrients into the soil in bands is another option. Placing the recommended nutrient rate in a concentrated band either on or below the soil surface (where soil moisture might improve nutrient uptake) can increase crop recovery of nutrients (figure 4–11). Subsurface placement depth ranges from 5 to 20 centimeters (2 to 8 inches), depending on the crop, nutrient source, and application equipment. Concentration of nutrients increases nutrient recovery by the plant, because the applied nutrients have less soil to interact with, increasing the nutrient concentration in the soil solution.

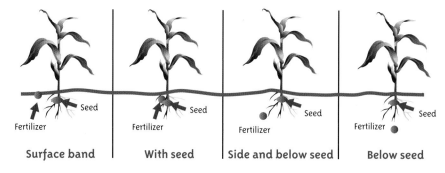

Figure 4–11. Options for band application of nutrients at or below the soil surface.

Figure 4–12. Foliar fertilization is when nutrients are applied directly to leaves.

Nutrients may also be applied directly to leaves to remedy obvious (visual) nutrient deficiencies or to prevent hidden hunger (not visually obvious) that can seriously impair crop yield or quality (figure 4–12). Leaf application can be an excellent supplement to soil-applied nutrients.

SUMMARY

Our own nutrition depends on soil. Understanding how different soil components influence nutrient supply to plants is essential to producing enough nutritious food to meet the growing global demand.

A farm field is not a closed system—it requires replenishment of the nutrients that leave whenever a crop is harvested, through crop residues or other sources. It is important to know how nutrients are conserved in soil and how soils can meet a crop's demand. Without sufficient nutrient supply, plants cannot reach their growth potential.

Plants tell us if they are nutrient stressed through deficiency symptoms that we can see. We can also collect a sample of soil to determine the nutrient supply through laboratory testing. Once nutrient needs are determined for a specific crop, the manager decides how and when the nutrients should be applied. Careful nutrient management is essential to avoid costs to the grower and the environment.

CREDITS

4–1 through 4–4, John Lambert; 4–5, SAGE, 2002—Center for Sustainability and the Global Environment, Inst. Environ. Studies, Univ. of Wisconsin, Madison (http://www.sage.wisc.edu/atlas/maps.php), with data from IGBP-DIS Global Soils Dataset (1998); 4–6, Johannes Lehmann; 4–7 and 4–8, Harold van Es; 4–9, John Lambert; 4–10, USDA-NRCS; 4–11, John Lambert; 4–12, Bob Nichols, USDA-NRCS. Other photos: nutrient deficiency, John Havlin; soil sampling, USDA-NRCS; leaf analysis, Stephen Ausmus, USDA-ARS. Chapter opener image, iStock.

Copyright © 2012. Soil Science Society of America, 5585 Guilford Rd., Madison, WI 53711-5801, USA. *Know Soil, Know Life*. David Lindbo, Deb A. Kozlowski, and Clay Robinson, Editors
doi:10.2136/2012.knowsoil.c4

GLOSSARY

ADSORB (ADSORPTION) Attachment of ions to mineral and organic matter surfaces in soil.

ANION Negatively charged ions (NO_3^-, Cl^-, SO_4^{-2}).

ANION EXCHANGE CAPACITY (AEC) Quantity of positive charges on surfaces of mineral and organic matter that attracts anions.

AMMONIA Form of nitrogen (NH_3) that converts in water to form plant-available ammonium (NH_4^+).

AMMONIFICATION The conversion of organic nitrogen to ammonium (NH_4^+) by the action of decomposers (bacteria).

ASSIMILATION Uptake or utilization of an element by an organism.

BASE SATURATION Proportion of the cation exchange capacity occupied with basic cations (Ca^{+2}, Mg^{+2}, K^+).

BUFFER Ability to the soil to replace ions in soil solution removed by plant uptake or other loss processes

CATION Positively charged ions (Ca^{+2}, H^+, Al^{+3}).

CATION EXCHANGE CAPACITY (CEC) Quantity of negative charges on surfaces of mineral and organic matter that attracts cations.

DEFICIENCY When a nutrient is not plentiful enough to allow for proper growth.

DENITRIFICATION In the absence of molecular oxygen (O_2), some bacteria can use oxidized nitrogen (nitrate and nitrite) as a terminal electron acceptor and produce dinitrogen gas (N_2).

DESORB (DESORPTION) Detachment of ions from mineral and organic matter surfaces to the soil solution.

DISSOLVE Transformation of a compound into its component cations and anions.

EXCHANGE SITES Charged sites on the surfaces of soil materials such as clay or organic matter that can store nutrients in ion form.

HIDDEN HUNGER Level of nutrient deficiency that reduces plant yield without observable visual deficiency symptoms.

IMMOBILIZATION Conversion of plant available nutrients to an organic form which is unavailable to plants.

ION EXCHANGE The interchange between an ion in solution and another ion in the boundary layer between the solution and surface of a charged material such as clay or organic matter.

LUXURY CONSUMPTION Ability of plants to absorb nutrients at levels above that needed for optimum plant growth and yield.

MACRONUTRIENT Essential elements (N, P, K, Ca, Mg, S) nutrient found in plants in the highest amounts.

MICRONUTRIENT Essential elements (Cl, Fe, B, Mn, Zn, Cu, Ni, Mo) nutrient found in plants in the lowest amounts. Also referred to as minor elements.

MINERALIZATION Conversion of an element from an organic to an inorganic form as a result of microbial activity.

MYCORRHIZA (PLURAL: MYCORRHIZAE) Literally means "fungus roots." the association, usually beneficial, of specific fungi with the roots of higher plants (adjective form is mycorrhizal).

NITROGEN CYCLE Describes the conversion of dinitrogen gas (N_2) from the atmosphere through microbes, plants, and animals to soil transformations and back to atmospheric dinitrogen gas. Important microbial processes include biological nitrogen fixation, mineralization, nitrification, and denitrification.

NITRIFICATION Biological oxidation (loss of electrons) of ammonium to nitrite and nitrate.

NUTRIENT An element essential to living organisms.

PRECIPITATE Specific cations and anions combine to form a compound.

RHIZOBIA General term to identify root-nodule bacteria, such as rhizobium and bradyrhizobium, that are active in biological nitrogen fixation.

VISUAL DEFICIENCY SYMPTOMS Low nutrient levels in plants that cause abnormal growth or discoloration.

CHAPTER 5

SOIL CLASSIFICATION, SOIL SURVEY, AND INTERPRETATIONS OF SOIL

So far this book has focused on the physical, biological, and chemical properties of soils and landscapes. This chapter ties these concepts together to show how the properties of soils are organized for use by scientists, engineers, farmers, students, or just about anyone who is curious about soils. For any subject in the natural world, scientists often begin with classification. This is true for plants, animals, insects, microorganisms, and for soils, too. Worldwide several soil classification systems are used, but here we focus on the system used by the U.S. Department of Agriculture Natural Resources Conservation Service (USDA-NRCS) called *Soil Taxonomy*.

Once soils are described and classified (chapter 2), the information is made more useful by showing where these soils are on the landscape—in other words, a map. Soil maps or soil surveys are similar to the range maps used for plant and animals species. Finally, once soil is classified and mapped, the data can be applied and interpreted to identify assorted land uses consistently.

SOIL CLASSIFICATION

Any type of classification system organizes knowledge into a structural, logical system that is easy to remember, enhances communication, and shows us the relationship between the subject being classified and other similar subjects. Classification should be flexible and help guide further research. Finally, and perhaps most importantly, it arranges items into groups to allow data to be interpreted easily and consistently. Soil Taxonomy strives to meet all these goals. As with many classification systems, Soil Taxonomy is hierarchical and follows a *dichotomous key,* so that any given soil can only be classified into one group.

CHAPTER AUTHORS

DAVID LINDBO
DOUG MALO
CLAY ROBINSON

In this sense, soil classification is much like the classification of plants. No two leaves on a tree are exactly alike, but they are distinctly different from those on a tree of a different species. Likewise, soils in a similar landscape that developed under the same CLORPT conditions (chapter 2) will have some characteristics in common but may have subtle differences (figure 5–1). The greater the differences in CLORPT, the more pronounced the differences in soils and classification. Comparing soils that share all the same or similar CLORPT factors is like comparing red oak (*Quercus rubrum* L.) to white oak (*Quercus alba* L.) (figure 5–2)—the differences are

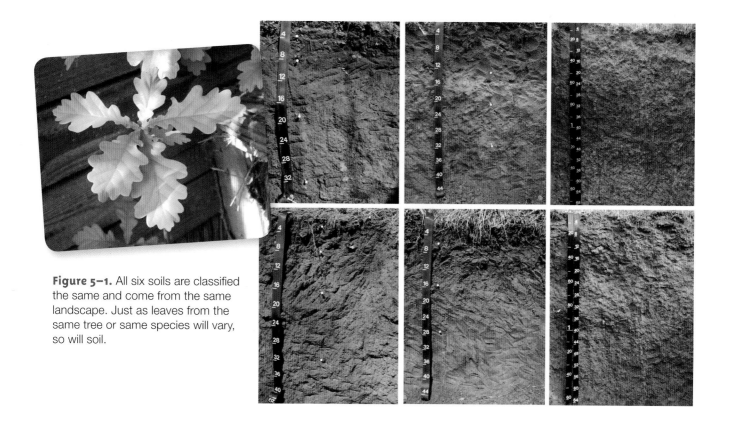

Figure 5–1. All six soils are classified the same and come from the same landscape. Just as leaves from the same tree or same species will vary, so will soil.

Figure 5–2. Both red and white oaks are classified as *Quercus* at the genus level, but one is *Quercus rubrum* and one is *Quercus alba*, two different species. The Appling (left) and Cecil (right) series are classified similarly but are different soil series. They have may similar properties and land uses. The primary difference is the color.

Figure 5–3. Continuing with the tree and soil comparison: Both pine and oak are trees—both the Hayesville (left) and Leon (right) are soil. They are taxonomically different and are classified differently—different species for the trees—different soil orders for the soils.

subtle. On the other hand, comparing soils that that don't share all or any CLORPT factors is like comparing oak and pine (figure 5–3)—the differences are profound.

Soil Taxonomy is composed of six levels: order, suborder, great group, subgroup, family, and series (figure 5–4, tables 5–1, 5–2). The system has been designed to classify any soil in the world (figure 5–5a, b, c) and can be complex. The taxonomic name of a soil contains a great deal of information about the nature and properties of that specific soil (see example in table 5–1). For the purposes of this book we will focus only on the highest (order) and lowest (series) levels.

Soil orders are the most general level of classification in soil taxonomy (table 5–2, figure 5–6a). Orders are similar to kingdoms in the Linnaeus system of classifying organisms. Each order is based on one important diagnostic feature such as permafrost for Gelisols or shrink–swell clays for Vertisols. The taxonomy follows a strict dichotomous sequence, so that once a soil is placed in an order it cannot be moved to another order. The key diagnostic feature for a given order is based on its significant effect on the land use or management of all soils in that order (table 5–2). The orders also represent different weathering intensities or degrees of formation (figure 5–6b).

Some orders are limited to specific regions or biomes (see chapter 7), while others are related to specific parent materials. In the United States, Mollisols and Aridisols dominate the subhumid to arid western regions, while Ultisols and Spodosols are most common in the humid eastern regions (figure 5–7). Alfisols are a transition between East and West. Ultisols are most common in the hot, humid areas of the southeastern United States. Spodosols can occur throughout the East (and in some areas of the Northwest) where the soils are sandy and the vegetation is dominated by conifers. Oxisols are confined to tropical environments in Hawaii and Puerto Rico. They are also extensive in other tropical areas worldwide. Gelisols are extensive where permafrost is present such as Alaska as well as northern Canada and Siberia.

Figure 5–4. The hierarchy of Soil Taxonomy.

12 ORDERS
70 SUBORDERS
330 GREAT GROUPS
1500+ SUBGROUPS
7000+ FAMILY
16800+ SERIES

Table 5–1. Examples of classification in the Linnaean taxonomy system and in Soil Taxonomy.

Linnaeus	
Kingdom	Plantae
(unranked)	Angiosperms
(unranked)	Eudicots
(unranked)	Rosids
Order	Fagales
Family	Fagaceae
Genus	Quercus
Section	Quercus
Species	Quercus alba

Soil Taxonomy		Example: Cecil Series (fine, kaolinitic thermic Typic Kanhapludult)
Order	major soil forming processes; diagnostic horizon, one dominant property	*ult*—Ultisol, highly weathered, acidic, clay accumulation
Suborder	genetically similar (moisture regime, organic matter content, parent material affects)	*ud*ult—Udic or humid moisture regime
Great group	diagnostic layers, base status, horizon expression, clay activity	*Kanhap*ludult—dominated by kaolinite clay in subsoil
Subgroup	central concept (Typic), intergrades and extragrades	*Typic* Kanhapludult—typical type of Kanhapludult
Family	land use related properties; texture, mineralogy, temperature	*Fine, kaolinitic, thermic*—clay content 35–60% in subsoil, kaolinite clay dominates in the clay fraction of the soil, thermic temperature regime
Series	kind and arrangement of horizons, specific properties of horizon	*Cecil*—named for location where first discovered. In this case Cecil County, Maryland

Table 5–2. The 12 soil orders, in the order in which they are classified in Soil Taxonomy.

Order		Brief description	Derivation	Mnemonicon, or syllable
Gelisols	el	has permafrost	Greek—*gelid*, very cold or frozen	**Gel**id
Histosol	ist	dominated by organic (O) horizons	Greek—*histos*, tissue	**Hist**ology
Spodosol	od	sandy and has an accumulation of organic matter, aluminum, and/or iron	Greek—*spodos*, wood ash Russian—*podzol*, ashy gray underneath	**Pod**zol
Andisol	and	formed in volcanic materials and/or has specific mineral properties related to density, phosphorus, and water holding	Japanese—*ando*, black	**And**o
Oxisol	ox	dominated by iron and aluminum oxides	French—*oxide*, oxidized	**Ox**ide
Vertisol	ert	dominated by shrink–swell clay	Latin—*verto*, turn	Inv**ert**
Aridisol	id	arid region, salt, carbonates, or gypsum accumulations	Latin—*aridus*, dry	**Arid**
Ultisol	ult	highly weathered, acidic, clay accumulation (e.g., kaolinite)	Latin—*ultimus*, last	**Ult**imate
Mollisol	oll	dominated by thick, dark-colored mineral surface, non-acid	Latin—*mollis*, soft	**Moll**ify
Alfisol	alf	moderately weathered, clay accumulation, slightly acidic	Nonsense historical term—Ped + al + fer = foot (in Latin) + aluminum + ferrous	Ped**alf**er
Inceptisol	ept	weakly developed, no clay accumulation (no Bt)	Latin—*inceptum*, beginning	Inc**ept**ion
Entisol	ent	relatively undeveloped (AC profile)	Recent, new	Rec**ent**

Histosols and Andisols are identified specifically by the soil's parent materials. Histosols are organic soils that occur where decomposition is slow and organic materials accumulate. This often happens at northern latitudes where cold temperatures slow decomposition, and also in broad, flat landscapes where water tables are shallow, such as the Everglades, and parts of eastern North Carolina and Virginia. Andisols form in volcanic materials and in the United States are limited to areas of the Pacific Northwest and Hawaii. They are common in much of Japan as well.

The interaction between parent material, climate, and landscape is significant in the formation of Vertisols. These soils occur in areas where the chemistry of the minerals in the parent materials, combined with other factors, contributes to the formation of *shrink–swell clays* as the soil forms. In the United States, extensive areas of Vertisols occur on the Gulf Coast from Texas to Alabama, and throughout the Mississippi Delta north to Illinois. Vertisols are also common in depressions and landscapes underlain by shale parent materials in the Great Plains. Vertisols are common in the Ethiopian Highlands and west-central India as well.

The last two orders, Entisols and Inceptisols, can occur anywhere. These are young, poorly developed, and slightly weathered soils. Entisols generally lack horizon development other than an A horizon. Inceptisols will have an A and B horizon but lack any of the other properties that would classify the soil as one of the other orders.

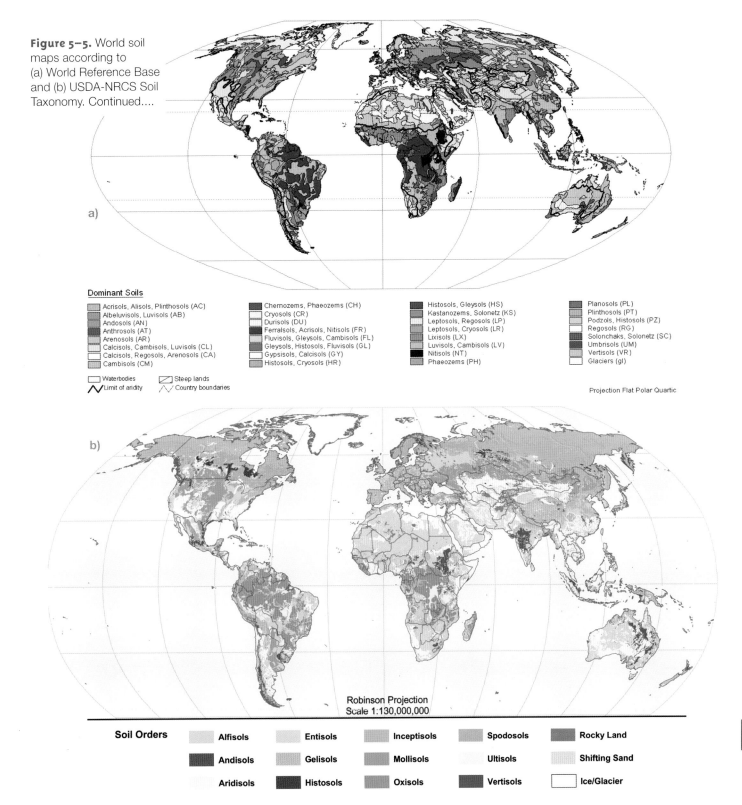

Figure 5–5. World soil maps according to (a) World Reference Base and (b) USDA-NRCS Soil Taxonomy. Continued....

Figure 5–5. Continued: (c) U.S. soil map.

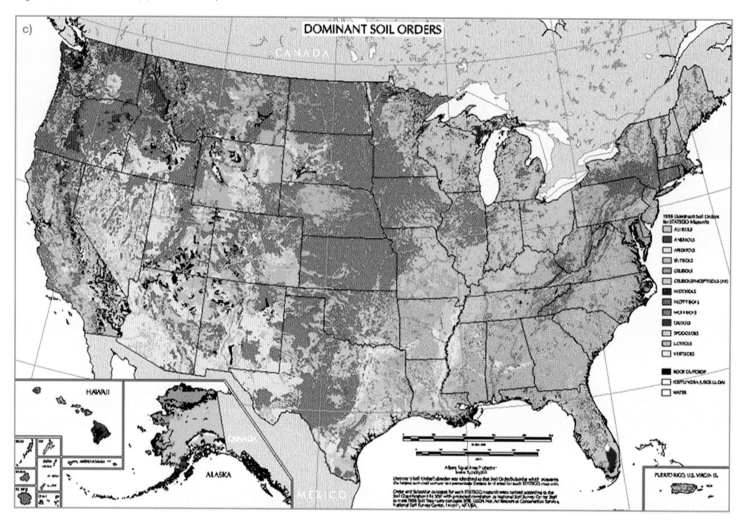

Naming Soils

The names of all soil orders contain the root "sol" (from the Latin *solum*, meaning soil). The prefixes in the name come from different languages and describe a specific property of the soil (table 5–2). For example, a soil that contains permafrost is by definition very cold. The name of the order for a permafrost soil is Gelisol, from the Greek word *gelid* for very cold. Each subsequent level of taxonomy is represented by another prefix. Each prefix has a specific meaning or properties associated with it (example in table 5–1).

sol

gelid

Figure 5–6. (a) The 12 soil orders of Soil Taxonomy. Continued...

a)

ALFISOLS

DOMINANT SUBORDERS
- Aqualfs
- Cryalfs
- Udalfs
- Ustalfs
- Xeralfs

ANDISOLS

DOMINANT SUBORDERS
- Aquands
- Cryands
- Torrands
- Udands
- Ustands
- Vitrands
- Xerands

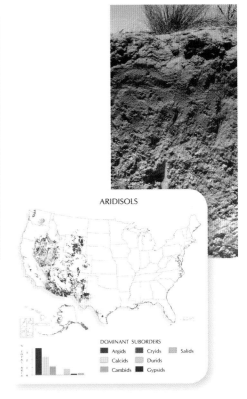

ARIDISOLS

DOMINANT SUBORDERS
- Argids
- Calcids
- Cambids
- Cryids
- Durids
- Gypsids
- Salids

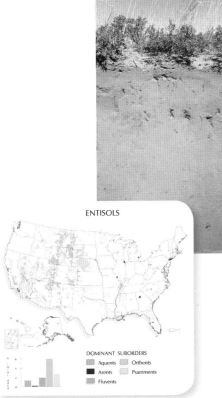

ENTISOLS

DOMINANT SUBORDERS
- Aquents
- Arents
- Fluvents
- Orthents
- Psamments

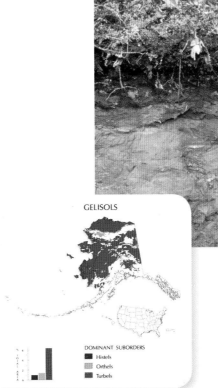

GELISOLS

DOMINANT SUBORDERS
- Histels
- Orthels
- Turbels

HISTOSOLS

DOMINANT SUBORDERS
- Fibrists
- Folists
- Hemists
- Saprists

Figure 5–6. Continued. (a) The 12 soil orders of Soil Taxonomy.

INCEPTISOLS
DOMINANT SUBORDERS
Anthrepts, Udepts, Aquepts, Ustepts, Cryepts, Xerepts

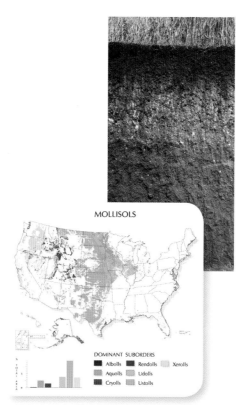
MOLLISOLS
DOMINANT SUBORDERS
Albolls, Rendolls, Xerolls, Aquolls, Udolls, Cryolls, Ustolls

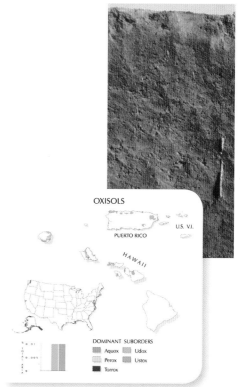
OXISOLS
DOMINANT SUBORDERS
Aquox, Udox, Perox, Ustox, Torrox

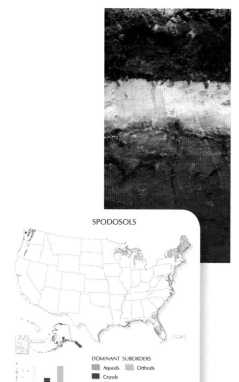
SPODOSOLS
DOMINANT SUBORDERS
Aquods, Orthods, Cryods, Humods

VERTISOLS
DOMINANT SUBORDERS
Aquerts, Uderts, Cryerts, Usterts, Torrerts, Xererts

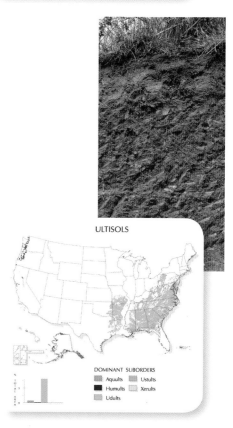
ULTISOLS
DOMINANT SUBORDERS
Aquults, Ustults, Humults, Xerults, Udults

Figure 5–6. Continued. (b) The orders represent a range of degrees of formation.

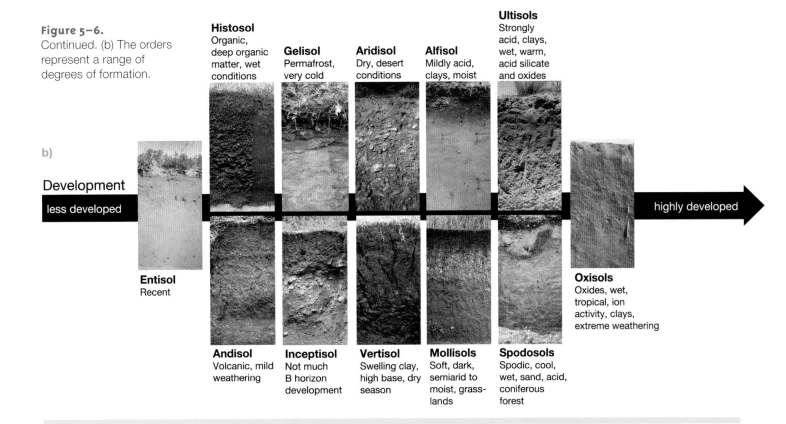

Figure 5–7. Transition: desert to prairie to forest. In terms of soils this is the Aridisol–Mollisol–Alfisol transition.

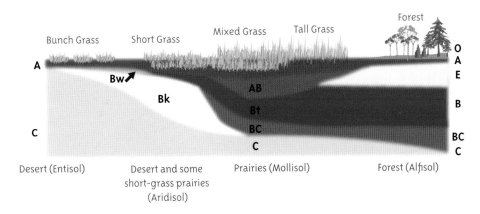

Once a soil's order has been identified, classification continues using a dichotomous key called the *Keys to Soil Taxonomy*. Identifying the order is relatively straightforward once the soil has been described and the key is followed, although routine laboratory analysis is sometimes needed. Because new soils are being discovered all the time, the keys are dynamic and are updated every few years to represent our ever-increasing knowledge base. Use of the keys further classifies the soil into a group, subgroup, family, and finally series, the lowest level of the classification. Series is analogous to the species level in Linnaeus classification. The sections that follow on soil survey and mapping require a basic understanding of the series concept.

A soil series is the same as the common name of the soil, much in the way that white oak is the common name for *Quercus alba* L. A *soil series* is defined based on a range of properties and is named for the location near where it was first identified. Scientists have identified more than 17,000 soil series in the United States alone.

For example, the Cecil series (figure 5–8) was first identified in Cecil County, Maryland. (It is also the state soil of North Carolina.) The Cecil series is an Ultisol and consists of very deep, well-drained (very deep water table), moderately permeable soils on upland ridges and side slopes. These soils formed in material weathered from felsic, igneous, and high-grade metamorphic rocks (see box in chapter 2). Slopes range from 0 to 25%. The Cecil series subsoil is dominantly clayey (>40%) and may be clay to sandy loam in the surface (in part related to the degree of erosion). The clay minerals present in the soil are mostly kaolinite.

If you know the series name of a soil, you can access a lot of information about the soil. The series concept and information contained within it can be used for many purposes and forms the basis for soil surveys and mapping.

SOIL SURVEYS AND MAPPING

Classifying and describing a soil gives us a lot of information about that particular soil; however, soils exist in a three-dimensional landscape, so scientists developed a method to convey this spatial information, known as *soil surveys*. Soil surveys are inventories of the soil resources in a geographic area and include detailed morphologic descriptions, physical and chemical properties of soil series, soil classification information, maps showing soil boundaries, and predictive interpretations for selected land uses.

An arm of the U.S. Department of Agriculture began to conduct soil surveys around the turn of the 20th century, concentrating on tobacco lands (figure 5–9). The goal was to investigate the relation of soils to climate, agriculture, and biology.

Early surveyors used the tools of the time including spades and augurs, plane tables, and alidades. They covered distances by motor vehicle, horse and buggy, and even bicycles specially adapted to travel on railroads; some even "rode the rails" as they traversed large distances in the survey efforts. The *Soil Survey* adopted the use of airplanes and aerial photography in the period between the world wars.

Soil Survey helped develop standards for the profession, including standards for describing soil color (see chapter 2). In the 1920s small vials of soil samples were used for field comparisons, and a standard soil color chart was eventually published. Concurrent with these efforts, Albert Munsell, a Massachusetts artist and art professor, developed the Munsell color system for use by industry and artists. In 1946, these color charts were adopted as the standard for soil color nearly 50 years after the Soil Survey began.

Figure 5–8. Cecil is one soil series. There are more than 17,000 soil series in the United States alone.

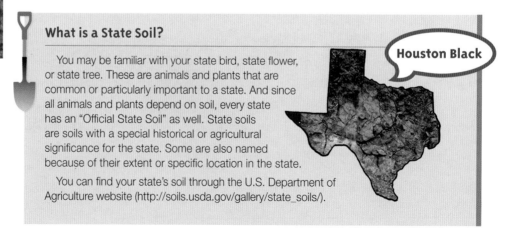

What is a State Soil?

You may be familiar with your state bird, state flower, or state tree. These are animals and plants that are common or particularly important to a state. And since all animals and plants depend on soil, every state has an "Official State Soil" as well. State soils are soils with a special historical or agricultural significance for the state. Some are also named because of their extent or specific location in the state.

You can find your state's soil through the U.S. Department of Agriculture website (http://soils.usda.gov/gallery/state_soils/).

Houston Black

Figure 5–9. An early published soil survey. An early soil map. Two members of the 1926 Polk County, Missouri Soil Survey, soil scientists H.V. Jordon and M.W. Beck, with their truck "Hilda."

Today soil scientists traverse the globe to study the world's soils. (from left) A soldier checks to make sure a soil pit is safe in Farah Province, Afghanistan as part of a USDA-NRCS Foreign Agricultural Service project to map soils and advise local residents on soil conservation. Soil scientists collected samples from the Rongbuk glacier of Mount Everest to study trace elements of pollutants transported thousands of miles to this remote, relatively untouched location. Bush planes are the only way to reach some remote locations, like parts of Alaska.

Technological advances now allow us to inventory and understand soil, one of our most basic agricultural and natural resources, in new and extremely useful ways. Along with the traditional spades and augers, soil surveyors today use backhoes, all-terrain vehicles, helicopters, pickups, and hydraulic probes to study soils all over the world (figure 5–9). Aerial photography used with stereoscopes is one of the most significant tools of the modern Soil Survey, as it greatly improves the base maps used in field mapping. Electromagnetic meters, ground penetrating radar, global positioning systems, digital cameras, and field computers are also commonly used.

SOIL SURVEY AND SOIL SURVEY METHODS

A soil survey includes description, characterization, classification, mapping, correlation, and interpretation of the soils of a county. The heart of a soil survey is the soil map showing the spatial distribution and variability of soils on the landscape. Soil scientists prepare the maps in the field using pits, core samples, or trenches to examine the soils. They outline the extent of different soils based on aerial photography, landscape positions, or landforms and vegetation. Knowledge of the interaction of the soil-forming factors in a landscape aid in locating soil differences and in mapping unit boundaries.

Besides showing where specific soils occur, a soil scientist must:

1. Describe the morphology (e.g., soil color, color patterns, horizons, horizon boundaries, field texture class, structure, rock fragments, redox features, roots, pores, effervescence, and structure).
2. Measure physical properties (e.g., percentages of sand, silt, and clay; moist consistence; stickiness; plasticity; bulk density; porosity; water-holding capacity; and hydraulic conductivity).
3. Determine chemical properties (e.g., pH, lime content, organic carbon levels, cation exchange capacity, salinity, and sodicity).
4. Describe site and profile characteristics (e.g., land form, parent material, water table depth, drainage class, percent slope, surface runoff, root restrictive layers, vegetation, rooting depth, GPS location, flooding, ponding, hill slope position, and vegetation).
5. Classify the soil (e.g., soil taxonomy, land capability classification, and engineering uses).
6. Interpret soil suitabilities and limitations for land use (e.g., crops, range, waste disposal, roads, buildings, wildlife habitat, recreational uses, gardens, lawns, and many others).

SOIL MAPPING

Soil mapping is a detailed descriptive process that begins with an understanding of the soil-landscape relations, field investigation, and cartography. Depending on the scale of mapping needed, the experience level of the soil scientist, and the complexity of the landscape, a soil scientist can map from 200 to more than 600 acres per day. Soil maps are checked and reviewed (correlated) on a regular basis for quality control.

LOCATING SOIL SURVEYS

One can obtain detailed soils information from published soil surveys in two ways: at a county office of the USDA-NRCS or online using the Web Soil Survey.

PRINT SURVEYS

Soil surveys from any county in the United States are organized in the same way. Each survey begins with a generic one-page explanation on how to use the survey, followed by sections that describe the nature of the county, its history and development, the physiography, relief and drainage, water supply, climate, and background on how the survey was made. A brief explanation of the map units in the survey is also included.

A soil *map unit* is a collection of map delineations spatially locating and identifying the presence of the same soil components (soil series). Soil mapping units can be *associations* (two or more soil series occurring together in a characteristic, repeating pattern but not as mixed as a complex); *consociations* (soil mapping unit is dominated by the named soil series while adjacent soils have properties and management requirements similar to the named soil); *complexes* (two or more soil series so intimately intermixed geographically that they cannot be separated at the mapping scale, with more mixing than for a soil association).

The survey contains two types of maps: the general soils map and the detailed soil maps. Users consult the general soil map for broad planning and for comparing the suitability of large sections of the county but not for detailed land-use planning, management at the farm or field scale, or for real-estate and land development. The detailed soil maps combined with the detailed map unit descriptions discuss the dominant soils, their properties, and their limitations and often include diagrams showing the relationships of soil series to landscapes (figure 5–10).

For information regarding specific areas, users consult the detailed soil maps. In most cases these maps are accurate to approximately five acres; they should not be used for detailed management or planning on smaller parcels of land. Detailed map unit descriptions give a non-technical soil description, and discuss the map unit's properties, limitations for various land uses, and related series that may be included in the map unit. Information is included about each soil series mapped in the county. Each soil series is described in detail including landscape, depth to water tables, parent materials, and variability.

Soil surveys also include a section on the use and management of soils. This

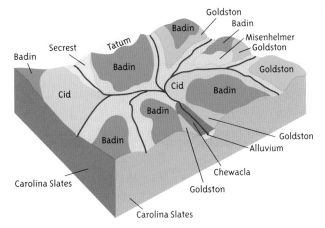

Figure 5–10. There is a relationship between the soil series and the landscape. What do you notice in this example from the Carolina Slate Belt?

land and soil data available online (see the appendix for some useful websites for producers and natural resource managers).

USES AND LIMITATIONS OF SOIL SURVEY INFORMATION

Soil surveys help us understand how soils differ and how they behave under various land management systems. The key soil properties determine recreation, crop production, range, water/erosion conservation, forestry, and engineering uses of the soil. Most of the soil maps in the printed Soil Survey and WSS were originally prepared at scales ranging from 1:10,000 to 1:48,000. As a result, what can be interpreted from the maps is limited by the mapping scale used.

The smallest area that can be shown on modern soil survey maps ranges from 0.4 to 4.9 hectares (1 to 12 acres), with 2.0 hectares (5 acres) being most common. Yet smaller areas (inclusions) may be extremely important in the management of a field. Most soil map unit descriptions

consists of a series of tables describing cropland, pasture and hay land, orchards, yields per acre, land capability classification, prime farmland, woodland or rangeland management and productivity, recreation, wildlife habitat, and land use issues, including building site development, sanitary facilities, construction materials, and water management. This section also rates the suitability for particular land uses ranging from "slight" limitations through "moderate" to "severe" limitations based on national standards set by USDA-NRCS. These ratings are used as a guideline and are not meant to be the last word in land-use planning or management.

ONLINE: THE WEB SOIL SURVEY

Today you can also obtain soil survey and other information online using the USDA-NRCS Web Soil Survey (WSS) (http://websoilsurvey.nrcs.usda.gov/app/HomePage.htm). A mobile device application is also available that works with iPhones and Android smart phones (http://casoilresource.lawr.ucdavis.edu/drupal/node/902). The WSS website (figure 5–11)

is regularly updated with new options, features, and data (table 5–3). WSS (version 2.3, released in 2011) provides modern detailed soil survey information in both visual and tabular electronic formats. In addition to the WSS, which is widely used by farmers, ranchers, natural resource managers, and planners, there are many other sources of

Figure 5–11. The Web Soil Survey home page.

Table 5-3. Some of the information available in soil surveys and from Web Soil Survey.

Tables on suitability and limitations	Soil properties and qualities
Building site development	soil chemical properties
Construction materials	soil erosion factors
Disaster recovery	soil physical properties
Land classifications	soil qualities and features
Land management	water features
Military operations	
Recreational development	
Sanitary facilities	
Vegetative productivity	
Waste management	
Water management	

Source	Category	Examples†
Soil suitability and limitation ratings	Building site development	Concrete/steel corrosion, dwellings with basements, lawns, local roads/streets, landscaping, golf fairways
	Construction materials	Gravel source, topsoil source, reclamation material, sand source, roadfill source
	Disaster recovery	Composting facility, composting media/cover, rubble/debris disposal, large animal disposal
	Land classifications	Land capability classification, Soil Taxonomy, forage/pasture suitability, ecological site, tree group
	Land management	Forest land use (haul roads, erosion hazards, dust, tree planting, soil rutting, equipment operability)
	Military operations	Bivouac areas, vehicle trafficability, landing zones, fighting positions
	Recreational development	Camp area, off-road motorcycles, paths and trails, picnic areas, playgrounds
	Sanitary facilities	Landfill site, landfill cover material, septic tank absorption fields, sewage lagoon, waste water use
	Vegetative productivity	Crop productivity ratings/yields for all major crops, forest productivity, range production
	Waste management	Manure and food-processing waste
	Water management	Pond creation, irrigation (general, sprinkler and flood)
Soil properties and qualities	Soil chemical properties	Soil pH, salinity (EC), sodicity (SAR), calcium carbonate content, cation exchange capacity
	Soil erosion factors	K factor, T factor, and wind erodibility group/index
	Soil physical properties	Available water holding capacity, bulk density, surface texture, percentages of organic matter, sand, and clay
	Soil qualities and features	Engineering properties, drainage class, frost action, parent materials, percent slope, root restriction depth
	Water features	Water table depth, flooding/ponding frequency

Table 5–3. Continued.

Source	Category	Examples†
Soil tabular data	Area of interest information	Map unit description, legend, soil survey area data summary
	Building site development	Dwellings with basements, lawns, local roads/streets, landscaping, golf, small commercial buildings
	Construction materials	Gravel source, topsoil source, reclamation material, sand source, roadfill source
	Land classifications	Land Capability Classification, Soil Taxonomy, tree/shrub suitability groups, hydric soils, prime farmland
	Land management	Forest land use (haul roads, erosion hazards, dust, tree planting, soil rutting, equipment operability
	Recreational development	Camp area, off-road motorcycles, paths and trails, picnic areas, playgrounds
	Sanitary facilities	Landfill site, landfill cover material, septic tank absorption fields, sewage lagoon, waste water use
	Soil chemical properties	Soil pH, salinity (EC), sodicity (SAR), calcium carbonate content, cation exchange capacity
	Soil erosion	K factor, T factor, wind erodibility group/index, attributes for soil loss calculations
	Soil physical properties	Available water holding capacity, bulk density, texture, percentage organic matter, engineering properties

† Note that not all soil properties and qualities ratings categories or all options within a category are listed in this table. Some items listed in this table may not be available in all counties and different items of local importance may be present.

include these features to let the user know they exist. For intensive management of areas smaller than 1 to 6 acres, a more detailed soil map is needed. GPS application methods and the use of modern soil survey data, along with yield and other management data, can give us an accurate picture of the soils we can expect to find in a given area.

INTERPRETATIONS

Collecting data on soils and developing maps is only the first step in relating soils to landscapes and land use. Once the data are collected interpretation begins. Soil data can be used for many things, from determining how much fertilizer is needed to grow a given crop to determining soil and site suitability for a shopping mall. All of these interpretations follow specific guidelines to ensure soil and soil data are interpreted uniformly. We focus here on three common interpretations of soil data: to identify land capability classes, hydric soils, and prime farmland.

LAND CAPABILITY CLASSES

Soil scientists use *land capability classes* to categorize soils according to their suitability primarily for agricultural uses. There are eight classes, designated by Roman numerals (I through VIII). Class I land (prime farmland) has the widest range of uses, the fewest limitations, and requires the fewest conservation measures to keep the soil productive (figure 5–12). Class VIII land has the narrowest range of uses and the most severe limitations.

Land classes I through IV are suitable for agricultural crops that require tillage. Class I land is suitable for very intensive use with no conservation practices except good management. Class IV land is suited for limited-tillage crop production and requires more intensive practices for soil and water conservation (see best management practices in chapter 6). Classes II and III fall in between (figures 5–13, 5–14, and 5–15).

Land classes V through VIII (figures 5–16, 5–17, 5–18, and 5–19) are generally best suited for perennial crops such as pasture, hay, and fruit trees as well as vineyards, range, or forest land. Class V requires no conservation practices, but should remain natural because of the land's unique characteristics. Examples of Class V land include high mountain valleys or swamps subject to flooding. Class VIII lands should be managed for environmental protection or recreational, wildlife, and aesthetic uses. Examples are rocky and stony land areas, tidal areas, and sand dunes.

Land capability classes can be further divided into four subclasses: *e, w, s,* and *c* (table 5–4). The subclasses indicate the most restrictive kind of conservation problem limiting use of the land, with "e" being the most important limitation and "c" the least important. Subclasses are designated by a lower-case letter after the capability class (for example, IIe, IIIw, IVs). Any letter can be applied to any of the classes (except Class I, which has no limitations) to clarify the major limitations. If a soil has both a severe erosion (e) limitation and a severe root zone limitation (s) only the "e" would be used.

In semiarid and arid regions, water is the most limiting factor to agricultural production. In these areas soil map units will receive two land capability classifications, the first as described above, and the second for irrigated land management.

Figure 5–12. Soils in Class I have few limitations that restrict their use and are considered prime farmland. These soils are the best in nearly all respects for both agricultural production and non-agricultural uses. They are deep (100 cm or more), well drained, and loamy textured with medium to high available water capacity, moderate permeability and none to moderate erosion. These soils are easily farmed and are among the most productive in the world. Slopes do not exceed 2%. Management should include best management practices suited for the site; at a minimum for agronomic use include annual soil testing and ideally cover crops.

Figure 5–13. (above and continued on next page) Examples of several Class II soils with different subclasses (e, s, w) represented. Soils in this class have limitations that reduce the choice of crop or plants and/or require moderate conservation practices. Although these soils are good and usually productive, some conditions (drainage class, soil depth, slow or rapid permeability, low available water capacity, or moderate soil depth) exclude them from Class I land. Only one limiting condition is needed to require a soil to be moved from Class I to Class II. Despite this many are included as Prime Farmland. Slopes range between 2 and 6%. Depth to water table may be the limiting factor with redoximorphic features (wetness features or w subclass) within 50 cm of the surface.

Class IIs

Class IIw

Figure 5–14. Soils in Class III have moderate to severe limitations that reduce the crop choice or require additional best management practices or both. Limitations similar to Class II soils may be present, but these limitations are more severe and restrictive. Class IIIe lands are strongly sloping (usually range from 6% to 10% and subject to moderate to severe erosion. When the water table is within 50 cm of the surface, drainage (w) is the limiting factor. Additionally, soils <50 cm thick to a restrictive zone or horizon (IIIs), presence of coarse-textured surface layers (IIIs), fine-textured subsoils with slow permeability (IIIw), or very low available water capacity (IIIs) are placed in this class.

Figure 5–15. (above and continued on next page) Soils in Class IV have very severe limitations that restrict the crop choice, require more intense management, or both. If erosion is severe, this land is good for occasional, well managed cultivation or should remain fallow (sod or grass rotation) for long periods. Slopes are between 10% and 15%. Areas with shallow water tables (IVw) require intensive drainage management for agriculture. Severely eroded sites with little to no topsoil or A horizons are included in this capability class, regardless of other conditions including slope. Intensive or extreme best management practices are required for sustained agricultural production on these soils. Rotation periods with hay, pasture, or cover crops of several years are required to minimize the erosion loss. Even with proper management crop failures may occur in this class.

Figure 5–16. Class V soils are considered to be unsuited for cultivation of row crops. Although soils in this class are nearly level and not subject to erosion, they are limited by excessive wetness from frequent flooding, ponding, or seepage or by rock outcrops, excessive stoniness or similar restrictions that render them unsuited for cultivation. However, they may be deep and have few limitations for pasture or forestry.

Figure 5–17. Soils in Class VI are generally unsuited for cultivation but can be used for pasture, woodland, or wildlife food and cover with some limitations. Common limitations include steep slopes (15–25%), a severe erosion hazard (e), effects of past erosion, and/or stoniness (s). Near constant shallow water tables (w) can also result in a soil being placed in Class VI or higher. However, with drainage a Class VIw or VIIw could be reclassified to a Class IVw.

Figure 5–18. The areas of Class VII soils are more limited than Class VI due to very steep slopes (>25%), very shallow soils, and very stony soils that occur on slopes (s). Their use is largely restricted to grazing, woodland, or wildlife. Proper and intensive best management practices can make productive pasture and woodland possible. Regardless of use special care is required to prevent excessive erosion (e). Soils with constant saturation near the surface (w) are classified into this class.

Figure 5–19. Areas in Class VIII are precluded from use in commercial agricultural or silvacultural production. Limitations restrict their use to recreation, water supply, wildlife, or aesthetic purposes. Tidal marshes with daily flooding, continuously ponded areas, and areas with >90% rock outcrop, stones or boulders, borrow pits, barren mines, dumps, urban areas and sandy beaches are all in this class.

Table 5–4. Land Capability Subclasses.

Subclass†	Description
e	water or wind erosion hazard
	water erosion subclasses are related to slope, texture, infiltration, and permeability
	conservation practices to control soil and water losses required
	sandy, loamy, or clayey textured surface horizons, loamy or clayey subsurfaces, that are shallow, moderately deep or deep, and on slopes above 2%, are in "e" water erosion subclasses
	wind erosion subclasses are primarily related to surface soil texture and landscape position
	wind erosion subclasses may be any texture class, and may occur on 0% slopes
w	wetness (high water table, flooding, ponding, or poor drainage) as the major limitation
	requires drainage, as indicated by low chroma colors in the matrix or as redoximorphic features
	soils that are located in floodplains and subject to flooding
s	variety of root zone or special soil limitations
	well-drained soils with sandy surface layers and sandy subsurface layers, with the total depth of sandy texture extending to >50 cm (20 inches) on any slope
	well-drained soils with sandy surface layers (50–100 cm thick, 20–40 inches) and loamy subsurface layers on any slope
	soils with more than 50% of the surface layer consisting of stony, cobbly, or gravelly materials
	very shallow soils (limiting layer at <30 cm, 12 inches) on all slopes
	soils with high salinity, high sodicity, high or low pH
c	extreme climatic conditions (cold, heat, or excessively dry due to lack of precipitation)
	cold extremes include soils at high elevations or latitudes
	deserts would be considered in the dry extreme but may occur in either cold or hot temperatures

† Note: Subclasses are listed in order of importance (i.e., "e" is more important than "w", which is more important than "s", etc.).

HYDRIC SOILS

Soil scientists often need to interpret soil data to identify hydric soils for the purpose of preserving wetlands under the Clean Water Act. Wetlands have important ecological functions, ranging from slowing water flow into rivers and lakes to improving water quality and providing wildlife habitat. The presence of hydric soils is only one characteristic of a wetland. Scientists also look for wetland hydrology and the predominance of hydrophytic (water-adapted) vegetation.

Hydric soils have developed under saturated and anaerobic conditions within 30 centimeters (12 inches) of the soil surface because of flooding, ponding, or a high water table for a portion of the growing season (see chapter 2). These soils can support hydrophytic vegetation.

Hydric soils are commonly identified based on their morphology (figure 5–20) or less commonly on the duration and frequency of saturation and anaerobic conditions measured over a long period in the field. Soil scientists typically use the hydric

Figure 5-20. Hydric soils (right) commonly have a low-chroma matrix caused by saturation and anaerobic conditions that cause redoximorphic features (see Chapter 2 for a discussion of redoximorphic features). The non-hydric soil (left) is typified by a high chroma matrix (above 2) in the upper 30 cm.

soil field indicators developed by USDA-NRCS to identify if the soil is a hydric soil. These indicators include organic soils (40 cm, 10 inches of organic soil material); organic surface layers (20–40 cm, 10–20 inches thick); gleyed or low chroma colors generally with the presence of redoximorphic features (chapter 2); and high organic matter in the surface or as streaks near the surface. More than 40 such field indicators, which are regionally based, are used to identify hydric soils. As with all forms of classification and interpretations, extensive training is required to properly use the indicators; most importantly, an accurate field description of the soil morphology is required.

PRIME FARMLAND

Interpreting soil data to identify prime farmland is of major importance because we depend on it for most of our food and fiber needs. As such, this identification is used to create an inventory of these important lands. It could also be used to identify these lands for protection from other uses.

Prime farmland comprises all Land Capability Class I land, and includes some Class II lands. The acreage of high-quality farmland is limited and deserves protection. All soil surveys of an area contain information on soils, if any, that are considered prime farmland. Prime farmland has the best combination of physical and chemical characteristics for producing food, feed, forage, fiber, and oilseed crops. Prime farmland is limited to areas that are in production or could be in production. Areas of prime farmland that have been developed for other purposes are no longer considered prime.

Scientists consider the soil and site properties when designating prime farmland. Favorable soil conditions include adequate fertility, acceptable acidity or alkalinity (pH), an acceptably low salt and sodium content, and few or no rocks. Prime farmland is permeable to water and air and is not excessively erodible or saturated with water for long periods. Prime farmland has enough moisture from precipitation or irrigation, favorable temperatures and growing season, is not frequently flooded during the growing season or is protected from flooding, and has gentle slopes ranging from 0% to 6%. (Class I land has 0% to 2% slopes.)

Prime farmland soils may be currently used as cropland, pasture (or rangeland), or forest land. Suburban or urban developments are excluded, as the land is not available for agriculture; however, the growing urban agriculture movement may change this. Public land is not designated as prime farmland since it is not available for agriculture. This removes large acreages of national forests and grasslands, national parks, military reservations, and state parks from the prime designation. Soils that have limitations (high water table, subject to flooding, or dry) may qualify if the limitations are properly managed and alleviated.

Many states have soils that are considered unique or important farmland soils. For example, the extensive Norfolk series in the southeastern United States has few management issues, commonly produces high crop yields, and thus is considered prime farmland. But such prime farmland is declining worldwide. It is lost to industrial and urban uses and, as the world population grows (see chapters 6 and 8), the loss of prime farmland results in more use of marginal lands for agriculture. These lands are more erodible, dry, and generally less productive. They are more difficult to farm and manage and often require greater inputs of energy to produce a crop, driving up the cost of production and the cost of food and fiber to consumers.

SUMMARY

Soils are a natural system that we have classified in a systematic way. Soil Taxonomy is the classification method used in the United States, although this system can be used worldwide. Soil classification looks at all aspects of the soil, from the landscape to the parent materials to the morphology, chemistry, and even the biology—in other words, it encompasses all aspects of CLORPT.

Classification is not simply an academic exercise for soil scientists; it has some important applications. First, classification forms the backbone of soil mapping. Soil maps allow farmers, developers, land use planners, and civil engineers, among others, to identify the soils in a given area. Second, once soils are identified, specific interpretations, applications, and limitations for the use and management of those soils can be established.

Land capability classes give a broad view of the agricultural uses and limitations of the land based on specific soil and site conditions. Likewise, prime farmland is identified from soil maps and soil classification. Hydric soils indicators are another example of how specific soil properties based on morphologic description and classification can identify soil for a particular land use. In the end, classification serves as a way to communicate information about a soil in a consistent fashion to soil scientists and land managers worldwide, with the ultimate goal of good stewardship of our soil resource.

CREDITS

5–1, imageafter.com, D. Lindbo; 5–2, J.S. Peterson, NPDT @ USDA-NRCS PLANTS Database, D. Lindbo; 5–3, imageafter.com, D. Lindbo; 5–5, FAO, USDA-NRCS; 5–6, D. Lindbo, USDA-NRCS; USDA-NRCS; 5–7, John Lambert; 5–8, USDA-NRCS, John Kelley; 5–9, USDA, USDA-NRCS, B. Yeo; 5–10, John Lambert, based on USDA survey; 5–11, USDA-NRCS Web Soil Survey; 5–12 through 5–19, USDA-NRCS, John Kelley, D. Lindbo, R. Vick; 5–20, D. Lindbo. Chapter opener image, iStock.

Copyright © 2012. Soil Science Society of America, 5585 Guilford Rd., Madison, WI 53711-5801, USA. *Know Soil, Know Life*. David Lindbo, Deb A. Kozlowski, and Clay Robinson, Editors
doi:10.2136/2012.knowsoil.c5

GLOSSARY

ALFISOLS Moderately leached soils often found in temperate forests—generally east of the Mississippi.

ANDISOLS Soils formed in volcanic ash—Pacific Northwest, Japan.

ARIDISOLS Desert soils—desert areas worldwide.

ASSOCIATION A map unit composed of two or more soil series occurring together in a characteristic, repeating pattern.

COMPLEX A map unit composed of two or more soil series intimately intermixed geographically that they cannot be separated at the mapping scale.

CONSOCIATION A map unit dominated by the named soil series while adjacent soils have properties and management requirements similar to the named soil.

DICHOTOMOUS KEY A key used to classify an item in which each stage presents two options, with a direction to another stage in the key, until the lowest level is reached.

ENTISOLS Soils with little or no morphological (horizon) development—beaches, sand dunes, and floodplains.

GELISOLS Soils with permafrost—tundra, Alaska, Siberia, Northern Canada.

HISTOSOLS Organic soils—very wet areas, parts of FL, MN, AK, MI, ME, NC.

INCEPTISOLS Weakly developed soils—almost anywhere.

LAND CAPABILITY CLASS One of the eight classes of land in the land capability classification of the usda-nrcs, distinguished according to the risk of land damage or the difficulty of land use.

MAP UNIT (i) a delineation identified by the same name in a soil survey that represent similar soil and landscape areas; (ii) a loose synonym for a delineation.

MOLLISOLS Grassland soils—the Great Plains, Russian steppes.

OXISOLS Very weathered soils of tropical and subtropical environments—Puerto Rico and Hawaii and other tropical areas such as Brazil and Southeast Asia.

SHRINK–SWELL CLAYS Clay minerals that expand greatly when wet or saturated, e.g., montmorillinite, bentonite.

SOIL MAP A map showing the distribution of soils or other soil map units in relation to the prominent physical and cultural features of the earth's surface. The following kinds of soil maps are recognized in the USA:

Detailed soil map: A soil map on which the boundaries are shown between all soils that are significant to potential use as field management systems.

Detailed reconnaissance soil map: A reconnaissance map on which some areas or features are shown in greater detail than usual, or than others.

Generalized soil map: A small-scale soil map which shows the general distribution of soils within a large area and thus in less detail than on a detailed soil map.

Reconnaissance soil map: A map showing the distribution of soils over a large area. The units shown are soil associations.

Schematic soil map: A soil map compiled from scant knowledge of the soils of new and undeveloped regions by the application of available information about the soil-formation factors of the area.

SOIL ORDER A group of soils in the broadest category. For example, in the 1938 classification system. The three soil orders were zonal soil, intrazonal soil, and azonal soil. In the 1975 there were 10 orders, whereas in the current USDA classification scheme there are 12 orders, differentiated by specific characteristics or properties: Alfisols, Andisols, Aridisols, Entisols, Gelisols, Histosols, Inceptisols, Mollisols, Oxisols, Spodosols, Ultisols, Vertisols. Orders are divided into suborders and the suborders are further divided into great groups.

SOIL SERIES The lowest category of u.s. system of soil taxonomy. A soil series is named for the area in which it was first mapped. A soil series is based on specific morphological, physical and chemical properties that make it unique. It is equivalent to the species level in Linnaean classification.

SOIL SURVEY (i) the systematic examination, description, classification, and mapping of soils in an area; (ii) the program of the national cooperative soil survey that includes describing, classifying, mapping, writing, and publishing information about soils of a specific area.

SOIL TAXONOMY U.S. Department of Agriculture–Natural Resource Conservation Service basic system of soil classification for making and interpreting soil surveys.

SPODOSOLS Acidic, sandy forest soils under conifers—sandy areas of the northeast to Minnesota, sandy areas of the Atlantic coastal plain.

ULTISOLS Acidic, strongly leached, older soils—common in the Southeastern United States and old piedmont landscapes worldwide.

VERTISOLS Clayey soils that swell when wet—parts of Texas west to the desert southwest and east through Alabama, Mississippi delta region, northern great plains and parts of California. Also common in parts of India and Ethiopia.

APPENDIX: ONLINE SOURCES OF SOILS AND NATURAL RESOURCES INFORMATION

AGRICULTURAL RESEARCH SERVICE, USDA
Home page, research results and projects
HTTP://WWW.ARS.USDA.GOV/MAIN/MAIN.HTM

AMERICAN FACT FINDER, U.S. CENSUS BUREAU
Source of population, housing, economic, and geographic data by town, county, or zip code area
HTTP://FACTFINDER2.CENSUS.GOV/FACES/NAV/JSF/PAGES/INDEX.XHTML

BUREAU OF LAND MANAGEMENT, USDOI
Home page, projects and activities
HTTP://WWW.BLM.GOV/WO/ST/EN.HTML

CALIFORNIA SOIL RESOURCE LAB.
Soil Survey Data
HTTP://CASOILRESOURCE.LAWR.UCDAVIS.EDU/DRUPAL/NODE/902

CANADA CENTRE FOR REMOTE SENSING
General remote sensing information
HTTP://CCRS.NRCAN.GC.CA/INDEX_E.PHP

CURRENT RESEARCH INFORMATION SYSTEM (CRIS)
Current agricultural research results and publications
HTTP://CRIS.NIFA.USDA.GOV/

EROS DATA CENTER, USGS
Home page, satellite and aerial images, research projects and programs
HTTP://EROS.USGS.GOV/

GOOGLE MAPS/ GOOGLE EARTH
Various maps in 2 and 3 dimensions
HTTP://MAPS.GOOGLE.COM/
HTTP://WWW.GOOGLE.COM/EARTH/INDEX.HTML

MAP STATS OF UNITED STATES
Federal statistics maps for state, county, and city
HTTP://WWW.FEDSTATS.GOV/QF/

NATIONAL AGRICULTURAL STATISTICS SERVICE
Agricultural statistics for state and county
HTTP://WWW.NASS.USDA.GOV/

NATIONAL INFORMATION MANAGEMENT AND SUPPORT SYSTEM (NIMSS)
Agricultural research activities and projects in the state, region, and nation
HTTP://NIMSS.UMD.EDU/

NATIONAL INSTITUTE OF FOOD AND AGRICULTURE (NIFA)
Home page and agricultural research information
HTTP://WWW.CSREES.USDA.GOV/

NATIONAL MAP VIEWER (USGS)
Various kinds and scales of US maps
HTTP://NATIONALMAP.GOV/VIEWERS.HTML

NRCS—FIELD OFFICE TECHNICAL GUIDE
Provide county specific scientific technical and reference information on soil, water, air, plant and animal conservation
HTTP://WWW.NRCS.USDA.GOV/TECHNICAL/EFOTG/

NRCS—HYDRIC SOILS
Hydric soils information
HTTP://SOILS.USDA.GOV/USE/HYDRIC/

NRCS—MAJOR LAND RESOURCE AREAS (MLRAS)
Physiography, geology, climate, water resources, soils, biological resources, and kinds of land use
HTTP://SOILS.USDA.GOV/SURVEY/GEOGRAPHY/MLRA/INDEX.HTML

NRCS—NATIONAL SOIL SURVEY HANDBOOK
Technical guide for soil survey projects and activities
HTTP://SOILS.USDA.GOV/TECHNICAL/HANDBOOK/

NRCS—NATIONAL RANGE AND PASTURE HANDBOOK
Procedures for the inventory, analysis, treatment, and management of grazing land resources
HTTP://WWW.GLTI.NRCS.USDA.GOV/TECHNICAL/PUBLICATIONS/NRPH.HTML

NRCS—NATIONAL CENTERS
National NRCS Centers (e.g., Water + Climate, Soil Survey, Agroforestry, and others)
HTTP://WWW.NRCS.USDA.GOV/ABOUT/ORGANIZATION/CENT_INST.HTML

NRCS—NATIONAL WATER AND CLIMATE CENTER
Climate and water conservation planning information
HTTP://WWW.WCC.NRCS.USDA.GOV/

NRCS—OFFICES/CENTERS
State and county office location and address information
HTTP://WWW.NRCS.USDA.GOV/ABOUT/ORGANIZATION/REGIONS.HTML

NRCS—OFFICIAL SOIL SERIES DESCRIPTIONS
Detailed, official soil series descriptions for soils in the United States
HTTP://SOILS.USDA.GOV/TECHNICAL/CLASSIFICATION/OSD/INDEX.HTML

NRCS—SOIL DATA MART
Soil physical, chemical, and characterization data
HTTP://SOILDATAMART.NRCS.USDA.GOV/

NRCS—SOIL EXTENT MAPPING TOOL
Map where named series are located
HTTP://WWW.CEI.PSU.EDU/SOILTOOL/SEMTOOL.HTML

NRCS—SOIL QUALITY
Soil quality definition, assessment, management, resources, and publications
HTTP://SOILS.USDA.GOV/SQI/

NRCS—SOIL SURVEY MANUAL
Soil Survey Manual Publication
HTTP://SOILS.USDA.GOV/TECHNICAL/MANUAL/

NRCS—SOILS
Home page, soil classification, lab data
HTTP://SOILS.USDA.GOV/

NRCS—TECHNICAL REFERENCES
Web site for manuals, technical guides, and references used by NRCS
HTTP://SOILS.USDA.GOV/TECHNICAL/

NOAA
Weather data, drought monitoring, current conditions
HTTP://WWW.WEATHER.GOV/

SERVICE CENTER LOCATOR (USDA)
Service Center locator and contact information
HTTP://OFFICES.SC.EGOV.USDA.GOV/LOCATOR/APP

SITE SPECIFIC MANAGEMENT GUIDE
Site specific management for agriculture
HTTP://WWW.IPNI.NET/E-CATALOG/SSMG/SSMG.HTM

SOIL ORDERS
Images of 12 soil orders
HTTP://SOILS.CALS.UIDAHO.EDU/SOILORDERS/

SOILWEB, ONLINE SOIL SURVEY BROWSER,
For Google Maps/Earth, real-time data app for iPhone and Android phones
HTTP://CASOILRESOURCE.LAWR.UCDAVIS.EDU/DRUPAL/BOOK/EXPORT/HTML/902

U.S. FOREST SERVICE
Home page
HTTP://WWW.FS.FED.US/

WEB SOIL SURVEY (WSS)
Detailed soil survey information
HTTP://WEBSOILSURVEY.NRCS.USDA.GOV/APP/HOMEPAGE.HTM

WORLD REFERENCE BASE FOR LAND RESOURCES
Soil classification and soil description for world, FAO
HTTP://WWW.FAO.ORG/NR/LAND/SOILS/SOIL/EN/

CHAPTER 6

ENVIRONMENTAL SCIENCE, SOIL CONSERVATION, AND LAND USE MANAGEMENT

John Wesley Powell (1834–1902) was a geologist, explorer of the American West, and a leading thinker on scientific subjects of his time. Powell understood the importance of soil to the growing nation, but he did not understand how fragile it is. In 1879 he wrote, "The soil is the one indestructible, immutable asset that the nation possesses. It is the one resource that cannot be exhausted, that cannot be used up."

Powell was incorrect on the matter of soil's durability. In the decades following his assessment, poor farming practices left the soil bare and susceptible to erosion. As a result, gullies appeared in fields, clouds of dust towered into the air, soil fertility and crop yields suffered, and air and water quality declined. The concept of sustainability was not a concern for the growing country.

Sustainability is existence maintained over time. A sustainable system is one where current and developing practices can continue without exhausting the natural resources on which that existence depends. In other words, resources are renewed or replenished rather than depleted. Sustainability promotes balance with the natural world, avoids damaging that balance, and encourages development in harmony with the natural world.

Many Native American tribes had an understanding of sustainability. Several required tribal leadership to consider the impact of new ways of harvesting food, for example, on the seventh generation to come, or almost 200 years into the future. Many native peoples were aware that current actions can have dramatic effects on future conditions, and they were concerned about the long-term costs of the short-term benefits the new settlers sought as they plowed up the prairies or hunted bison indiscriminately.

In this chapter, we examine several natural and human threats to sustained soil productivity, which may become irreversible if not mitigated. To reduce agricultural and environmental degradation and losses, steps must be taken to replace poor farming and development practices with those that reduce human effects and remain sustainable in the long run.

CHAPTER AUTHORS
CLAY ROBINSON
WALE ADEWUNMI
DAVID LINDBO
BIANCA MOEBIUS-CLUNE

Such practices are referred to as *best management practices* (BMPs). These are activities that landowners and managers (whether urban, suburban, or rural, in forestry or agriculture) can use to help conserve soil and water resources. BMPs are proven to reduce erosion and pollution and improve water and environmental quality. Management practices can either address the source of the problem, the outcome of the problem, or both. We will discuss some common problems, and how BMPs can be used to manage them (summarized in table 6–1).

As we learned in chapter 1, although about 38% of Earth's land surface is agricultural, only 11% is *arable* land, capable of sustained production of the food and fiber needed to feed and clothe the approximately 7 billion people on Earth. Most arable soils currently are used to grow crops. The rest of the soils are too steep, shallow, hot, cold, wet, dry, or have chemical limitations that limit their potential to provide for an ever-growing populace. Many factors, both natural and human-induced, may reduce the productivity of soil in natural and agricultural ecosystems. We examine these factors in the next sections.

Table 6–1. Best management practices (BMPs) for agricultural soils.

BMP	Description	General use	Specific or scale of use	Notes
Conservation tillage • no-till • strip-till • minimum-till • ridge till	• Requires crop residue or mulch left on the surface • Residue cover should be evenly distributed over at least 30% of the soil surface • No-till means there is no traditional cultivation, e.g., plowing and harrowing • May include cover crops (grasses and/or legumes) that are grown, then killed with herbicides just before the new crop is planted • Stems and roots hold the soil together • Crop residue on the surface reduces evaporation and erosion, adds organic matter • Decreasing number of tillage operations decreases compaction	• Erosion control—water, wind • Water conservation • Compaction mitigation • Decrease runoff • Increase infiltration • Improve soil structure • Improve tilth and water quality	Agriculture: small and large scale	Often used in combination with contour farming, strip cropping, cover crops and other BMPs. Many producers throughout the United States have adopted conservation tillage practices. United Nations Food and Agriculture Organization promotes conservation tillage around the world.
Contour farming	• Tillage on a nearly level grade following the curves or contour of the terrain • Creates zones on the landscape that slow the flow of water downhill and divert it along the contour within the rows • Reduces plant nutrient losses, improves surface water quality, and reduces soil erosion • Reduces fuel consumption since equipment travels along on the same elevations	• Erosion control—water • Mitigate nutrient losses • Decreases runoff, increases infiltration • Improve water quality	Agriculture: small and large scale	Generally practiced on croplands with slopes greater than 5%.
Cover crops	• Planting vegetative cover on bare fields, generally done in the autumn but in some cases is done in the spring to prepare a field for a late season crop • Protect the bare soil by limiting erosion and providing organic material to the soil • Used with any crop including vineyards but are most commonly used with low residue crops such as, soybean, tobacco, cotton, peanuts, and vegetables • May be harvested (rye or oats) or plowed under (vetch, clover) to provide organic material and nutrients (nitrogen in particular) for the soil, microbes and plant use	• Erosion control—wind, water • Provide nutrients	Agriculture: small and large scale Construction and mine sites, preparation for revegetation	Often combined with conservation tillage and crop rotation. May be used as cover for a no-till BMP. Most commonly used in humid regions. Sometimes used with irrigation.
Crop residue management	• Cutting up and leaving the unharvested crop residue on top of the ground to provide ground cover • Provides mulch to reduce the erosive force of rain drop impact and slow runoff, allows more water to soak into the soil • Reduces wind and water erosion by preventing particle *entrainment* • The organic material added also reduces soil compaction	• Erosion control—water, wind • Organic matter management and carbon sequestration	Agriculture: small and large scale	Minimum, no-till, strip-till, and ridge-till are the central component. Many producers throughout the United States have adopted conservation tillage practices. United Nations Food and Agriculture Organization promotes conservation tillage around the world.

BMP	Description	General use	Specific or scale of use	Notes
Crop rotation	• Systematically alternates crops over several years on the same field • May include grasses (small grains like oats, wheat, and rye), legumes (alfalfa, clover or vetch) and field crops (corn, sorghum, cotton, soybeans) and several vegetable families. • Root systems of grasses help hold the soil in place • Leaves or blades reduce the erosive force of raindrops slowing down particle entrainment and erosion. • Reduces soil and water losses, maintains or improves physical, chemical, and biological conditions of the soil • Enhances nutrient cycling, especially when legumes are included, as symbiotic nitrogen-fixing organisms live in nodules on roots • Alternating crop type (grasses vs. broadleaves) and growing season (cool vs. warm) helps control weed, insect, and disease pests	• Erosion control—water, wind • Provide nutrients • Pest control • Organic matter management	Agriculture: small and large scale	The specific rotation is based upon the soil and field limitations and is designed to maximize soil and water conservation. Often used with contour farming, with strip cropping, grassed waterways, filter strips, diversions, field borders, no-till, sod, or grass rotations, and IPM. More diverse rotations are possible in humid regions.
Diversion	• Managed channel that redirects surface flow around or out of the field to a zone where the water can infiltrate and be filtered of pollutants such as sediment • Berms used on construction and mine sites to redirect water away from the active work zone • Many uses to direct water in cities and industrial areas, redirecting water from parking lots, away from roads, developments, commercial areas, etc.	• Erosion control—water • Water quality	Agriculture: small and large scale Construction, mine sites Municipal and industrial	Often combined with grassed waterways and filter strips. Used in all regions with sufficient slopes and precipitation events to generate runoff.
Field borders, vegetative buffers	• Grass or permanent vegetation on the edge of cropped field • Improve water quality by reducing, removing, or filtering sediment, organics, nutrients, pesticides, and other contaminants from surface water flow • Create an area for equipment travel and turnaround	• Water quality • Compaction mitigation	Agriculture: small and large scale	Often combined with many agricultural BMPs. Management for wildlife, such as quail, turkey, rabbits, and predators is another benefit of these zones. Practiced more extensively in humid regions with sufficient precipitation to sustain the border.
Filter strips	• Permanent vegetation (generally grasses) planted above and around ponds, streams, rivers, lakes, constructed channels, and other sensitive areas • Filter sediment and other pollutants from rainwater runoff • Close-growing grass slows the movement of water, increasing infiltration while trapping sediments	• Water quality	Agriculture: small and large scale Active and reclaimed mine sites Municipalities and industrial sites	Often combined with diversions, terraces and grassed waterways. Sometimes wildlife plantings are used in conjunction with filter strips. Often used in riparian zones.
Grassed waterways	• Natural or constructed channels lined with grass carry runoff water to a discharge point • The channel has slopes that are easy to mow and is not so steep as to allow erosion • Grass traps sediment, decreasing sediment loading to surface water	• Water quality	Agriculture: small and large scale Construction, mine and industrial sites Municipalities, parks	Commonly used with contour farming, terraces, diversions, and filter strips. Commonly used in areas with sloping soils and precipitation events that generate runoff. Grassed waterways are used where gully erosion is a problem.
Integrated pest management (IPM)	• Reduces the amount of pesticide used through a series of inter-related practices: chemical, mechanical, biological, cultural, and regulatory	• Pest control • Water quality	Agriculture, plant and animal: small and large scale	Often combined with other agricultural BMPs. Widely used throughout the United States, Canada and Europe.

BMP	Description	General use	Specific or scale of use	Notes
Riparian buffer	• Vegetation located adjacent to streams, lakes, ponds, and wetlands • Designed and maintained riparian buffers intercept and trap sediment, nutrients, pesticides, and other materials in surface and subsurface flow thus protecting and improving water quality • Vegetation in the riparian buffer generally includes native grasses, shrubs, and trees	• Water quality	Agriculture: small and large scale Construction, industrial and mine sites Municipalities Public lands	Often combined with filter strips, grassed waterways, diversions. National and State Parks (camping, hiking, etc.).
Soil testing (annual)	• Identifies the amounts of essential plant nutrients in the soil to determine fertilizer and lime rates • Lack of sufficient nutrients can reduce yields and result in poor plant cover of the surface • Too much of some nutrients, elements and/or salts can limit plant growth and yield, and increase potential for nutrients to enter ground or surface water	• Nutrient management • Water quality	Agriculture: small and large scale Municipalities: parks, golf courses, playing fields	Often combined with other BMPs. Primarily used in developed countries (due to cost of testing).
Strip cropping	• Crops in strips or bands to reduce erosion • Done on the contour, this practice is called contour strip cropping • Done without regard to the slope, this is called field strip cropping • Can reduce water erosion by 65 to 75% on 3 to 8% slopes • Alternating strips of rye or wheat with cotton can decrease wind erosion	• Erosion control— water, wind	Agriculture: small and large scale	Often combined with conservation tillage, crop residue management, crop rotation, terraces, and grassed waterways. Strip cropping is used around the world, primarily on sloping lands, but on flat lands for wind erosion control.
Terrace	• Construction of embankment ridges of soil along the contour to control runoff and break up the slope • Surface water running down the slope is collected and slowed, decreasing erosion potential • Spreads the water along the contour where it infiltrates into the soil • When terraces have slight slope to remove water from field, this is called a graded terrace • When terraces are designed with a sloped area and a flat area (to receive runoff), this is called a conservation bench terrace	• Erosion control—water • Water conservation	Agriculture: small and large scale	Often combined with conservation tillage, crop residue management, and other BMPs. The water may be channeled into a grassed waterway. Used throughout the world. Some terraces in China, Japan, and in the Andes Mountains are hundreds to thousands of years old.
Water control structure	• Small adjustable dams (wood or metal) placed in drainage ditches to manage water table levels • Also used in irrigation ditches to control water distribution • Allow for quick adjustments to the outlet to allow water flow out in times of high rainfall or storage in drier times, or to provide water to an irrigation channel • Limits nutrient leaching in areas with high water tables and artificial drainage • Can force denitrifcation to occur by favoring anaerobic conditions in the soil • Pesticides are held back and allowed to break down through microbial activity	• Water management • Nutrient control	Agriculture: small and large scale Irrigation Construction and mine sites	In flat landscapes in humid regions with high water tables. In irrigated regions, controlling water distribution and drainage.
Windbreaks	• Long rows of trees and shrubs planted perpendicular to the prevailing wind direction adjacent to open areas and plowed fields • Windbreaks reduce wind speed, decreasing detachment and transportation, and trap dust, reducing soil carried away by wind • Trap dust reducing the amount of soil carried away by wind • Most effective on small fields. Windspeed is reduced downwind for a distance about ten times the tree height	• Erosion control—Wind • Climate moderation (wind chill factor, atmospheric demand, etc.)	Plant and animal agriculture: small and large scale Shelter livestock and homes	Commonly used in areas with flat landscapes, sandy to silty soils, and high prevailing wind. The tree species vary from region to region based on climatic condition.
Specialized BMPs	• Detention basins hold water allowing sediments to settle so water that overflows the basin carries less soil • Silt fence reduces off-site sediment transportation • Erosion blankets are often used on steep grades (like highway overpasses) to reduce erosion • Hydro-mulch sprays a slurry of water and hay or other material to cover the surface and limit erosion • Revegetation is used to establish permanent vegetation after the project is complete	• Erosion control—water • Water quality	Construction and mine	Used in combination for best results.

NATURAL PROCESSES AFFECTING SOIL DEGRADATION

Natural processes that affect soil in both natural and agricultural ecosystems include erosion, acidification, desertification, and salinization. In natural ecosystems, these processes generally occur on geologic time scales, but human activity often acclerates them in agricultural ecosystems.

EROSION

Erosion is a threat to sustained agricultural production—and therefore to sustaining human populations and cultures. *Erosion* occurs when soil particles are detached, transported, and deposited (figure 6–1). These soil particles may or may not leave the field or area from which they were eroded. We learned in chapter 2 that water, wind, ice (glaciers), and gravity are involved in transporting soil materials, and these are also the active agents in erosion. All are at work in natural ecosystems. We focus here on water and wind.

Although erosion occurs in natural ecosystems, the rate of soil formation (see chapter 2) in humid and semiarid regions is approximately equal to the rate of erosion, so a somewhat constant amount of soil remains in place. This is not true of deserts, in which many soils have lost the entire A horizon, leaving bare surfaces, stone pavement (gravel left on the surface after all the sand, silt, and clay particles are removed) (figure 6–2), pedestaled plants (figure 6–3), and/or gullied landscapes.

Because of the way they are managed, agricultural ecosystems may experience accelerated wind and water erosion, so that soil erodes faster than it forms. If more soil erodes than forms, the amount of available soil decreases over time, threatening long-term sustainability. There are many examples of civilizations declining because they did not properly manage agricultural soils (see chapter 8).

Historical (and many current) methods of agricultural production include turning (inverting) the soil, which buries crop residues and breaks up soil aggregates, leaving a bare, uniform surface in which to plant seeds (figure 6–4). Bare soil aggregates are exposed to the energy in raindrops and wind. Such conditions promote accelerated water and wind erosion.

Erosion removes the finer particles from the surface. These particles take nutrients and organic matter with them, decreasing the fertility and productive potential of the soil. The effect on agriculture is loss of fertile ground, which necessitates increased use of fertilizer to sustain crop yield, which in turn increases the cost of food production.

Figure 6–2. Stone pavement results when gravel is left on the surface after all the fines are removed.

Figure 6–3. Pedestaled plants where roots hold the soil, but erosion removes the soil around the plant, resulting in a pedestal shape.

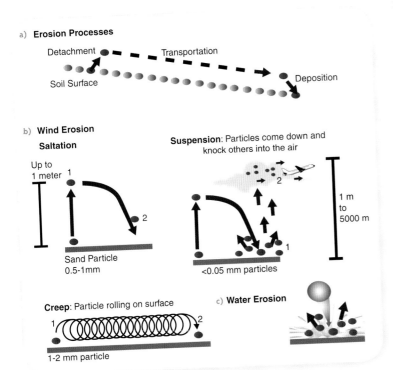

Figure 6–1. (a) All erosion processes involve detachment, transportation, and deposition of particles. (b) Three types of wind erosion—saltation, suspension, creep. (c) One form of water erosion is splash erosion, when raindrops detach soil particles. Flowing water also detaches and moves soil, in the forms of sheet, rill, and gully erosion.

Figure 6–4. Conventional inversion tillage involves cutting into the soil with a plow and "folding over" the soil, leaving no residues on the soil surface. The ridges are beds about 1 meter apart (39 inches) where the crop will be planted.

trees). Since 1982, the combined wind and water erosion on agricultural cropland in the United States has been reduced by more than 43%, from 3.03 to 1.73 billion tons of soil annually (figure 6–6).

Not all areas of the United States have been equally effective at reducing erosion. The Southern Plains region has two to three times the erosion rate of most other U.S. regions (figure 6–7), mostly due to the production of dryland cotton on sandy soils. Cotton plants leave little residue after harvest, and many cotton farmers still use traditional tillage methods.

WORLDWIDE, A STAGGERING 26.4 BILLION TONS OF SOIL IS LOST EACH YEAR. THAT RATE OF LOSS IS 10 TIMES FASTER THAN SOIL IS BEING REPLENISHED. THIS IS CLEARLY NOT SUSTAINABLE.

At its worst, from about the 1950s through the 1970s, erosion from agricultural cropland in some areas exceeded 20 megagrams (metric tons) per hectare per year (8.9 tons per acre per year). About 2% of the top 15 centimeters (6 inches) of topsoil was being lost annually from some farms, decreasing agricultural productivity of those croplands by as much as 30%. However, several factors in the last 50 years have begun to reverse those trends. Better tillage systems have been developed that leave more crop residues on the surface (figure 6–5). Some of the most erodible soils were taken out of crop production and planted to permanent vegetation (grass or

Figure 6–5. In this no-till field, you can see the new corn crop growing among the residue of last year's wheat crop stubble. This field is managed in a wheat–corn–fallow rotation. During the fallow year no crop is planted, which allows the soil to store water for next year's crop. The no-till practice increases water storage significantly by decreasing evapotranspiration and increasing infiltration.

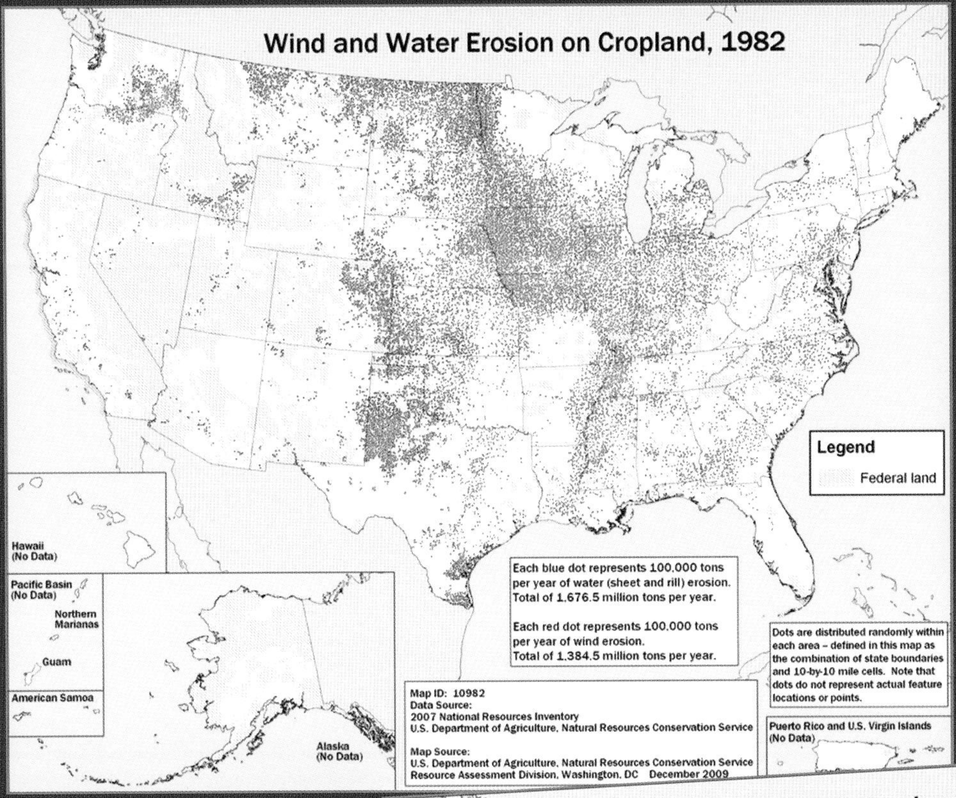

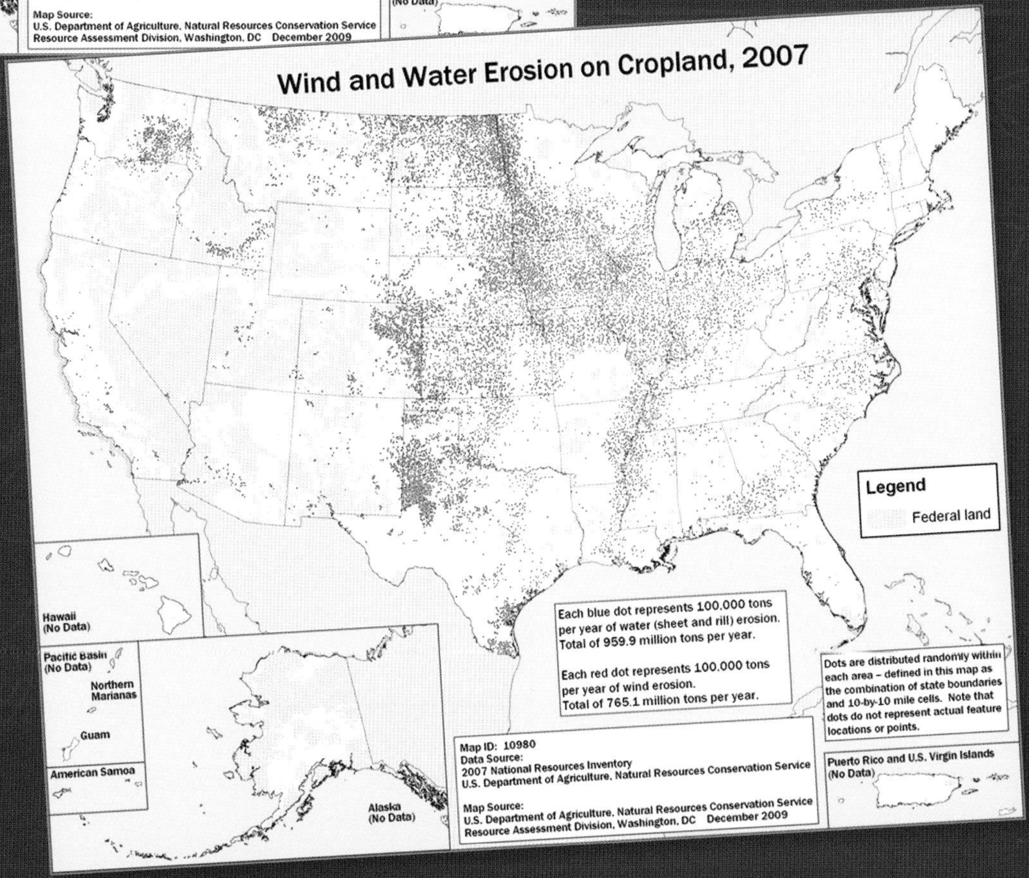

Figure 6–6. We've made progress in our management of erosion. Each dot represents 100,000 tons of erosion. The red dots are wind erosion, and the blue dots are water erosion. Compare the 1982 and 2007 maps—do you think in the future there can be a map with no dots?

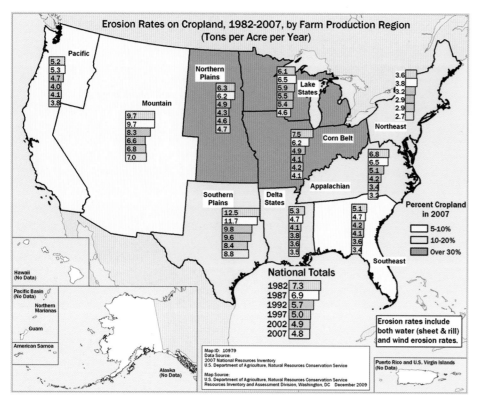

Figure 6–7. Erosion rates by region in the United States.

Rivers, lakes, estuaries, bays, and water bodies receive runoff that carries sediment (displaced soil) from construction sites, agricultural land, and urbanized areas. The Red Rivers in Texas and Oklahoma and North Dakota and Minnesota, the Yellow River in China, and the "Big Muddy" (nickname of the Missouri River) were named for the color of the sediment loads they carried (figure 6–8). Such sediments fertilize floodplains and deltas, but they are also a prominent source of water pollution. Eroded topsoil is detached from land and transported to water bodies where it is detrimental. Increased sediment in water degrades water quality and discolors water, making it less attractive for swimming, fishing, and other recreational endeavors. Sediments deposited in lakes and reservoirs decrease depth, thereby decreasing the volume of available water. Sediments fill up river and lake beds where some organisms take refuge and fish lay eggs. The shallow water spreads over a larger surface area, increasing evaporation. In addition, sediment clogs hydroelectric turbines and increases the cost of drinking

WATER EROSION AND BEST MANAGEMENT PRACTICES

You can see the long-term effects of water erosion in the shape of landscapes in many regions of the United States and around the world. Older, unglaciated landscapes such as those of Missouri and southern Iowa are typically more deeply incised, with more stream channels, rivers, and gently rolling topography, than are the relatively flat plains of the Dakotas and Minnesota. Rivers meander and change course as stream banks erode. Weathering and water erosion create beaches, which continue to change due to tidal erosion. Beaches can also be devastated by erosion from hurricanes and tsunamis.

Figure 6–8. Is the water "dirty" because it's brown? A more exact explanation is that runoff carries sediment to the river.

Figure 6–9. Rill erosion at the bottom of a slope in a wheat field.

water purification. For all these reasons, it is important to employ BMPs to decrease water erosion on managed landscapes.

Water erosion is caused by two detaching forces: raindrop impact and flowing water. Raindrops falling on soil surfaces may destroy soil aggregates and detach particles. Raindrops also may transport particles short distances (splash erosion). Flowing water transports particles, but may detach them as well. The erosive capacity of raindrops is due to their kinetic energy, which increases with the square of their velocity. Raindrops are small, but move fast and so have a lot of energy when they strike bare soil surfaces. In flowing water, both the mass of water and its velocity affect the ability of the water to detach and transport particles. The size of the particles transported by water increases with the kinetic energy of the water.

Three types of water erosion can occur: sheet, rill, and gully. *Sheet erosion* is difficult to see; essentially a uniform layer of soil is removed from a field or area. *Rill erosion* begins as water flowing across the soil surface concentrates into small rivulets, creating small channels (figure 6–9). *Gullies* are exaggerated rills, with deeper and wider cuts (figure 6–10). Gullies are a big problem when they form on croplands, since tractors and tillage implements cannot get across them.

Water control methods attempt to decrease the kinetic energy of the water by limiting soil detachment, decreasing transportation, and encouraging deposition.

The most effective way to control erosion is to keep the soil surface covered, either with growing plants or residues from past crops (figure 6–11). Plants and residues absorb the impact of raindrops, decreasing the kinetic energy when the raindrops reach the soil. Stems and leaves of plants or residue mats slow the rate of water flow across the surface, decreasing the kinetic energy and allowing more time for the water to infiltrate the soil,

Figure 6–10. Gullies are like rills on steroids; they are deeper and wider. These gullies began as wheel tracks on a dirt road.

Figure 6–11. Irrigated cotton planted after wheat. Wheat is planted in the fall to protect the soil from wind erosion and then "terminated" with herbicide prior to planting cotton.

Figure 6–12. The USDA seal features a plow as a symbol of agricultural production.

thus decreasing runoff, detachment, and transportation and increasing deposition. To control water erosion, farmers employ a variety of BMPs (table 6–1): reduced tillage, conservation tillage, residue management, contour plowing, strip cropping, cover crops, crop rotations, diversions, terraces, grassed waterways, water control structures, and buffer strips (also called vegetative filter strips).

Tillage is part of the fabric of agricultural crop production. A plow is included on the seal of the U.S. Department of Agriculture (figure 6–12). Historically, farmers employed tillage for three purposes: to prepare seedbeds, incorporate manure and crop residues, and control weeds. Before land was used for crop production, it was typically cleared of native vegetation (trees and/or grass) and then plowed. More recently in the semiarid Great Plains, tillage was found to have a negative role in water conservation; farmers learned that decreasing tillage and leaving residues on the soil surface decreased evaporation and increased the amount of water stored in the soil. Now we better understand the connection between tillage, soil condition, and erosion and are developing methods to decrease the erosion and improve soil condition.

Reduced tillage simply decreases the amount of soil disturbed during each tillage event by decreasing the number or intensity of tillage events. Reducing soil disturbance and the number of tractor trips over fields are both BMPs that leave more crop residues on the soil surface. Inversion tillage methods are extensive, for example, using moldboard and disc plows buries up to 90% of the crop residue on the first trip through a field after harvest. Each additional trip buries more residues, so the soil surface becomes bare after only one or two tillage events. This is especially true for low and fragile residue crops like cotton and seed legumes such as soybeans, peanuts, beans, and peas. Fragile residues are easily broken down by any tillage. Rice, corn, and wheat produce more residue, and the residues do not break down as rapidly as fragile crop residues. *Residue management* and *conservation tillage* are closely related, as both focus on having sufficient residue cover on the soil surface to control erosion when the next crop is planted.

One of the earliest forms of reduced tillage was sweep tillage. Long before larger versions of sweep plows were used in the United States, human-powered sweep plows (figure 6–13a) were used in China. Terraces in Gansu Province are covered with a stone mulch to limit water erosion and evaporation (figure 6–13b). The sweep controls weeds with limited mixing of the rocks and soil. A plow with one-meter-wide sweeps (figure 6–14) disturbs less soil, and leaves 80% to 90% of the plant residues on the soil surface (figure 6–15).

The ultimate reduced tillage system is called *no-till*. True no-till systems disturb the soil only when a new crop is being planted (figure 6–16).

Despite the benefits, many farmers worldwide have not adopted tillage

John Deere

In 1837, an innovative Illinois blacksmith, John Deere, revolutionized tillage when he developed a polished stainless-steel plow from a second-hand saw blade. It replaced the iron or steel plows that were prone to rusting and breaking in the prairie soils. Plowing land to plant crops became faster and easier. The new plow was used to invert the soil of millions of acres to incorporate residues and begin seedbed preparation. At that time, no one had an inkling of the long-term, adverse implications of this new technology.

Jethro Tull

In the early 1730s, an English agriculturalist, Jethro Tull, wrote *Horse-hoeing Husbandry*, in which he introduced his new horse-drawn seed drill. At the time planting primarily involved broadcasting (throwing or spreading) seeds across the soil surface. The drill was an important innovation in crop production as it placed seeds directly in the soil. But Tull also demonstrated a misunderstanding of plant nutrition, suggesting that the soil should be pulverized so that plant roots could eat it (just as we puree baby food today). It would be almost a century later before scientists began to understand plant nutrients and growth, and that "pulverizing" soil damages soil structure and increases compaction and erosion.

Figure 6–13. (a) This human-powered sweep plow is pulled under the stone mulch, killing the weeds but minimizing soil disturbance and mixing of the stones into the soil. (b) Peach trees, wheat, and canola grow on terraces mulched with stone.

systems that conserve residues, treating them like wastes to be removed rather than a valuable resource to manage. In some regions, farmers still call residues "trash." That attitude may be partially responsible for the practice of burning crop residues before plowing (figure 6–17). Cultural and societal norms may prevent farmers from adopting new tillage management practices that break from what everyone has always done. For example, farmers adopting minimum tillage practices may be considered "lazy."

Several BMPs are designed to reduce runoff on sloping soils. By reducing surface runoff, soil erosion and nutrient losses also are reduced. For example, *contour cultivation* involves tilling and planting across (perpendicular to) the slope along contour lines rather than up and down the hill (figure 6–18). Ridges created by contouring form barriers that slow surface water flow, increase water infiltration into the soil, and

Figure 6–14. A sweep plow blade, about 30 centimeters wide.

Figure 6–16. A farmer plants wheat with a 27-meter-wide (90 feet) no-till drill into existing residue.

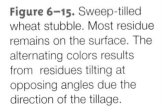

Figure 6–15. Sweep-tilled wheat stubble. Most residue remains on the surface. The alternating colors results from residues tilting at opposing angles due the direction of the tillage.

Figure 6–17. (a) Burning wheat residue to make plowing easier and (b) conventional plowing. Can you find the cloud of "dust," which is movement of soil fines by wind erosion? The direction of tillage is up and down the slope—how might that also move the soil?

Figure 6–18. Contour strip cropping, alternating bands of corn (golden stripes) and alfalfa (green stripes) in a field.

decrease erosion. *Strip cropping* involves growing alternate strips of annual (such as corn) and perennial (such as hay) crops on contours to similarly slow down water movement (figure 6–18).

Farmers often use *cover crops* to cover bare fields. Among other benefits, cover crops provide a canopy, like an umbrella, to limit the impact of raindrops on the soil surface and to slow runoff. They also may be used as cover in a no-till BMP.

Crop rotation involves alternating different crop families systematically over several years on the same field. The best rotations alternate broadleaf crops (beans and cotton) and grass crops (like wheat and corn). Similarly, warm-season crops (like corn, cotton, and soybeans) can be effectively alternated with cool-season crops (like wheat, oat, barley, and vetch). The specific crops in a rotation depend on whether the farm's main purpose is to grow grain crops, vegetable crops, or to produce milk or meat animals.

Crop rotation that includes longer-season crops and especially perennial crops (which grow over multiple years) reduces soil losses by providing protection for a greater fraction of the year. Good crop rotations also maintain or improve physical, chemical, and biological conditions of the soil. A specific rotation can be designed based on the soil to maximize soil and water conservation, and to break pest cycles.

Other BMPs are designed to move runoff away from, or off the field slowly while minimizing erosion potential. For example, *diversions* are berms or channels cut across hillsides to divert surface flow away from, or out of, the field to where the water can infiltrate and sediments can be deposited. A diversion is most useful in areas where upslope water runs onto the field, increasing the risk of rill and gully formation. Diverting the water away from the field reduces this risk.

Terraces are berms created within a field on gentle gradients across the slope to gather runoff water and carry it off the field. The spacing of terraces depends on the complexity and gradient of the slope. Terraces are used most effectively with grassed waterways. A *grassed waterway* (figure 6–19) (in the center, right side of photo) is built on a slight grade (typically 1% to 2%), and is designed to channel and slow runoff from agricultural lands, increasing infiltration of water and deposition of eroded sediments. It includes a stable outlet for water to prevent further erosion. Unlike a ditch, the channel is planted to perennial vegetation.

Figure 6–19. A grassed waterway is visible in this field as a thin green strip running through the brown, recently plowed, bare soil. Two grassed waterways come in from the right and top, join, then leave the field on the left side.

Soils and Pest Control

As we learned in chapter 3, pesticides can harm the soil ecosystem. Their use also can affect water quality. Farmers therefore often use *integrated pest management* (IPM) to reduce the use of pesticides and keep pests under control.

IPM employs five methods—chemical, mechanical, cultural, biological, and regulatory—to control weeds, insects, nematodes, rodents, fungi, and other disease-causing pathogens. All these organisms spend at least part of their life cycle in the soil, so controlling them affects the soil. IPM works to find the optimum combinations of all five tools to achieve the best pest control possible while limiting adverse effects to the ecosystem and the environment.

One approach to controlling pests is using chemicals specially formulated for that purpose, such as herbicides, insecticides, nematicides, rodenticides, and fungicides. Some of these chemicals may have adverse effects on the environment, especially water quality and non-target organisms. For example, using an insecticide to kill aphids that are damaging wheat plants in a field or roses in a garden also kills beneficial insects such as ladybird beetles. Controlling ants with insecticides has an unintended consequence of endangering horned lizards, since ants are the lizards' primary food. Some chemicals adsorb strongly to soil particles and may move to surface water with eroded soil. Other chemicals are not adsorbed, and like nitrates, may leach through the soil into groundwater.

The oldest pest control methods are mechanical and cultural. Mechanical control includes pulling or hoeing weeds in gardens and small fields or tillage in larger areas. Tillage buries plant residues that could otherwise house insects and pathogens. Another method uses fire to remove residues and kill weed seeds, insects, and pathogens. Tillage and fire must be used with care, as they leave the soil surface bare and increase the potential for soil to be lost to erosion. Some insects can be physically removed from plant leaves and stems, whether as adults, nymphs, larvae, or eggs. Cultural control methods include crop rotations which alternate types and seasons of crops to break pest cycles. These BMPs also are used to reduce erosion and improve soil physical properties.

Some pests have natural enemies. Biological control methods use these natural enemies to control pests. Insects such as ladybird beetles, dragonflies, and many wasps are natural predators of insect pests. Some bacteria strains limit the reproduction of some caterpillars that damage crop plants. Some weevils and mites feed on specific weeds, such as field bindweed. Sometimes the biological control agent can itself become a pest, so the release of new insects or other organisms as a control is regulated.

Other regulations specify the conditions where certain management practices are used. For boll weevil control in cotton, the planting dates, monitoring and cultural practices, and insecticide use and timing are all set by regulation. Regulatory methods also include alternating the types of insecticides and herbicides used, to limit the target pest from developing resistance to the chemical.

Cotton boll weevil on a cotton boll. Weevils lay eggs in flowers and bolls. The larvae eat their way out, decreasing yield and quality of cotton fibers.

Water control structures are small adjustable dams placed in drainage canals to manage water in flat landscapes. They allow quick adjustments to water flow depending on precipitation and soil conditions. Similar structures are used in canal irrigation systems.

Field borders, or vegetative *buffers,* are planted on the edges of agricultural lands (as on the left edge of the photo in figure 6–19) to limit runoff. Filter strips are similar in design, but are used on the margins of streams, rivers, ponds, and lakes to limit runoff and eroded sediments from leaving fields or construction sites. As little as five meters (16.5 feet) of vegetated buffer can decrease the amount of sediment leaving a field or entering surface waters by 80% or more.

WIND EROSION AND BEST MANAGEMENT PRACTICES

The effects of wind erosion were most dramatically illustrated by the Dust Bowl of the 1930s (see chapter 8), possibly the worst environmental disaster in the history of the United States. The Homestead Act, which encouraged settlement of the Great Plains, required some plowing and crop planting. Other government policies in the early 1900s combined with favorable wheat prices encouraged settlers to plant more than 100 million acres in dryland crops by the 1930s. Inversion tillage systems left no crop residues on the soil surface (figure 6–20). When rains did not come, all the conditions were right for a disaster.

Winds came and lifted the bare, dry soils (figure 6–21). Dust storms towering more than 3,000 meters (10,000 feet) darkened the skies two to three times a week. Croplands were destroyed. People and animals contracted dust pneumonia; many died. The Dust Bowl ended only when rains came after a decade-long drought.

A shorter, but more severe, drought occurred a couple decades later, resulting in the "Filthy Fifties." The development of *irrigation* and better residue management methods decreased its impact. But wind

Figure 6–20. A bare soil surface where wind erosion detached and transported fine particles away from the beds and deposited coarser sand particles in the furrows, or interrows. Conventional tillage left no residues on the soil surface to protect it from wind erosion.

Figure 6–21. "Black Sunday" occurred on April 14,1935, seen here as it approached Pampa, TX.

Figure 6–22. Wind erosion has buried this fence only a few years after it was built.

erosion remains a concern in the western Great Plains (figure 6–22) and elsewhere. Wind erosion damages millions of hectares of land in the United States annually, especially in the Great Plains.

Wind erosion is worse in the western Great Plains than in the eastern parts for several reasons. The west receives less precipitation, so there is less vegetation on the soil surface. The landscapes generally are flatter, with few hills and trees to slow the wind at the soil surface. The soils also tend to have less organic matter, and therefore less "glue" to hold aggregates together, making the soils more susceptible to wind erosion.

Although all soils can be moved by wind, sands are the particles that begin moving first. Silt particles may be carried for a few hundred kilometers, and clays for thousands, but sand sized particles start the process. Because of their rounded shapes, they get lifted into the air by the same physical mechanism that allows a plane to fly. The moving sand particles detach silt and clay particles, knocking them high into the air, and into the wind stream to be transported great distances. Sand particles are heavier and are moved a few centimeters to a few kilometers.

Wind erosion is common in many deserts that do not have vegetation to protect the soil surface. Some deserts have stone pavements on the surface because all the finer particles have been removed by wind erosion. Soil particles eroded from the Sahara desert are actually carried across the Atlantic and deposited in the Caribbean Islands and northern South America, increasing the fertility of those fragile, weathered tropical soils.

As in water erosion, kinetic energy is the primary force in wind erosion, and the energy of the wind increases with the square of the velocity. There are three types of wind erosion: *surface creep, saltation,* and *suspension* (figure 6–1). When the wind speed at the soil surface exceeds 21 kilometers per hour (about 13 miles per hour), sand particles begin to creep, or roll, along the surface. As the wind speed increases, sand particles begin to "jump" off the soil surface and into the air, which is saltation. The relatively large particles fall back to the surface. Driven by the wind, they gain kinetic energy. When they hit the surface, they dislodge other particles, knocking them into the air. These include smaller sand particles, as well as silt and clay. These particles can be lifted high into the atmosphere, which is suspension. Suspended particles can be transported long distances until the wind speed decreases enough that the particles drop and are deposited. Many soils form in these wind-deposited sediments, including those in the Palouse region in Washington and Idaho, much of the central and southern Great Plains, southern and western Iowa and Illinois, and many areas in western China along the Yellow River.

The main principles for controlling wind erosion are similar to those for controlling water erosion: protect the surface and reduce the energy. A covered surface protects soil from wind erosion; crop residues and growing plants reduce wind speed at the surface. Minimum and no-till systems also limit wind erosion (figure 6–23). For example, sweep plows and sandfighters (special implements designed for that purpose) rough up the soil surface, creating turbulence and decreasing wind speed at the surface. Tillage is a short-term solution, however; in sandy soils, the next rain will smooth the surface again. Windbreak plantings are commonly used around country homes, livestock facilities, and businesses but have little impact on agricultural land, since they only decrease wind speed for a distance of about 10 times the height of the trees in the windbreak.

Figure 6–23. No-till cotton in wheat residue.

Figure 6–24. Areas of the world most vulnerable to human-induced desertification.

DESERTIFICATION

Desertification is the extreme degradation of productive land in arid and semi-arid regions. Some 10 to 20% of the world's dry lands are already degraded, and land area in more than 100 countries is at risk of desertification. Affected regions are found throughout the world, including the western United States, Australia, sub-Saharan Africa, parts of the Middle East, China, and Central and South America (figure 6–24).

Desertification is a natural process that is associated with global climate change. With time, as the climate changes, even forests may become deserts. The Petrified Forest National Park, for example, now sits in the middle of the Arizona desert. Its huge fossilized logs are the remains of a prehistoric forest.

Desertification can also be brought on by improper management practices. Grasslands adjacent to deserts, such as the Sahara in Africa or the Gobi in Asia, are usually characterized by low precipitation amounts that vary greatly from year to year. Although they have low productivity and *carrying capacity*, such grasslands are typically used as rangeland for grazing cattle, sheep, goats, or other livestock. Putting more animals on an area than it can support results in overgrazing: too much of the grass is harvested, and the perennial grasses decrease in vigor. Continued overgrazing will cause the death of these grasses, resulting in degraded rangeland conditions (figure 6–25). The lack of soil cover increases the potential for wind and water erosion of both soil particles and grass seeds. Without sufficient precipitation, the perennial grasses do not recover or reseed themselves, and shrubs and annual grasses more characteristic of desert vegetation encroach. The land becomes less productive, and overgrazing often continues, exacerbating the conversion. China and Africa face the most severe challenges, as 30% of the lands in each are subject to desertification.

Range management BMPs include controlling the stocking rate and grazing intensity, keeping livestock out of sensitive areas, and providing livestock with alternative locations for water, and in some regions, shade. In open rangeland, adequate vegetative cover should be maintained to prevent accelerated erosion by wind and rain. Controlled grazing addresses the timing and duration in which animals have access to a pasture. One method includes

Figure 6–25. Degraded rangeland with crusted soils and sparse annual vegetation.

Figure 6–26. Degraded rangeland improvement project. The taller grass is present where a ripping plow was used on contours across the slope to allow water infiltration from precipitation events with runoff. A ripping plow is a deep shank that was pulled through the ground, ripping the soil perpendicular to the slope at 3-m (10-ft) intervals. Grass was seeded, and in some cases, compost was added upslope of the rip. A small bit of runoff occurs upslope and is caught by the compost and the slot in the ripped soil. The little bit of extra water is sufficient to alow the perennial grasses to become established again. With proper stocking rates (i.e., not too many animals) and grazing management, the vegetation spreads upslope, gradually closing the gap between the rips.

dividing larger pastures into paddocks, and rotating animals so that they can graze in one paddock while grass in another has time to regrow.

Proper range management and stocking rates will not reverse the effects of global climate change; however, in the short term, they can decrease soil degradation and slow the advance of desertification. Researchers have developed some techniques to reverse the effects of range degradation. Controlled burns are used to remove old growth and control certain invasive species. In other areas, invasive prickly pear and mesquite trees are sprayed with chemicals or physically removed to return the land to the climax vegetation of native grasses. When degradation is severe, more drastic measures are required (figure 6–26).

ACIDIFICATION

As we learned in chapter 4, the soil solution holds many minerals, some of which are more soluble than others. The soluble forms are ions that move through the soil with water. The ions that affect the soil's pH are cations. Calcium, magnesium, potassium, and sodium are basic cations, and maintain soil pH from neutral to alkaline. Hydrogen, aluminum, iron, and manganese are acidic cations.

In subhumid and humid regions, at times during the year the amount of precipitation exceeds what the soil can hold and plants can use. Some of the water moves through the plant root zone, carrying the soluble cations with it below the soil, and often into groundwater. *Acidification* occurs when the basic cations leach from the soil, leaving the acidic cations behind. The pH decreases as the soil becomes more acid. This natural process can be accelerated by the application of certain fertilizers. Producers managing acid farmlands regularly apply lime (calcium carbonate or similar minerals) to mitigate the acid in the soil.

SALINIZATION

Arid and semiarid lands, as in the southwestern United States, the Middle East, Australia, China, and parts of India, are subject to another devastating challenge to their soil's productivity: *salinization*, or the buildup of salts. The natural growth potential of plants in these regions is limited by the lack of water, even though the soils can be very fertile. Consequently, farmers and governments invest in irrigation technologies. Irrigation can increase crop yields by up to 10 times, but without proper management irrigation can turn soil into a salty, degraded mess. About 20% of the total amount of irrigated land on the planet has already been damaged by salinization (figure 6–27), and each year more than 1.5 million hectares (5,800 square

Figure 6–27. In hot and dry regions, fields need to be irrigated to grow crops successfully. Adding water to the soil increases evaporation. Sometimes saline irrigation water is used. In other cases water from saturated zones or saline aquifers moves upward by capillary action to replace the water being evaporated, bringing with it the salts naturally present in the soil and rocks below. When this water evaporates from the surface, the salts are left behind.

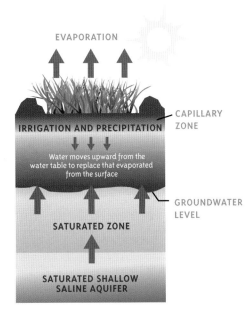

miles) of once arable land becomes unproductive from salinization.

Like ocean water, fresh water also contains salt, though in lesser amounts. Salinization occurs when the water evaporates from soil in arid and semiarid regions. The previously dissolved salts get left behind and begin to accumulate in the soil. Since these areas are dry, very little rainfall is available to wash out (leach) these salts.

As water evaporates from the soil surface, a gradient is generated that pulls water upward, if any is available, much like sipping a drink through a straw. When the water table is near the surface, it is easier for water to move to the surface and carry salts with it. When the water table is lower (deeper below the surface), it is more difficult for salts to move to the surface. In the Near East, 29% of irrigated lands have salinity problems, and 34% have been drained to lower the water table.

Although it can occur when groundwater is used, salinization is most commonly associated with irrigation from rivers. Irrigated fields are often in lowlands near rivers. River water in dry regions already contains elevated concentrations of salts, and high evaporative demand leads to some of these salts accumulating in the surface soil (figure 6–28). The unused fraction of the applied irrigation water leaches salts from the soil and carries them back into the river, further increasing salt content in river water that is then applied to fields downstream. As this occurs, the decreasing water quantity in the river becomes more saline (more salty). When this water is used to irrigate farmland, salts accumulate in part because of the increasing salt concentration in the water.

In addition, poorly managed or excessive irrigation also leads to rising groundwater tables in lowlands. As brackish (saline) groundwater rises, *capillary action* (see chapter 2) pulls this water up to the surface, where the water evaporates or is used by plants, leaving the salts behind to accumulate at the soil surface (figure 6–28). Remediation of these soils involves periodic leaching (adding more water than plants need). This additional water flushes some of the salts below the root zone. However, care must be taken not to add so much water that the groundwater table rises, where the groundwater table is close to the surface. In such cases a drainage system to remove excess water from the root zone may be required.

High salt concentrations are harmful to growing plants. Salts "compete" with plants for water. Plants generally have high concentrations of ions inside their roots to be able to absorb water, but as the salt concentration in the soil solution increases, plants are less able to take advantage of the osmotic gradient to pull water into their roots. Plants vary in tolerance to salts, but too much salt will prevent most plant species from obtaining sufficient water, even from a wet soil. As a result salinized land is then no longer productive. (See chapter 8 for a discussion of historical and current examples of salinization.)

A related problem sometimes encountered in arid regions is *sodification*, an increase in the sodium levels in the soil. Sodium is less of a plant problem, although high concentrations of sodium (also referred to as sodicity) may interfere with uptake of other nutrients and may be toxic to some tree species. The major impact of sodium is on the soil.

Figure 6–28. Salt accumulated on soil surface and base of plants.

Sodium effects begin at much lower concentrations than those that affect plant growth. Sodium is a dispersing or deflocculating agent. For this reason it is used in many detergents, as it pushes things apart. Other cations, especially calcium and magnesium, are flocculating cations, and pull particles together to form aggregates. When the concentration of sodium is high relative to calcium and magnesium, the sodium cations dominate the soil cation exchange sites and essentially push the particles apart, especially clays. The soil structure degrades, and the dispersed particles clog the pores in the soil.

Such a degraded soil structure creates surface crusts, limiting water infiltration, increasing runoff, and reducing plant productivity. Remediation of sodic soils requires the appropriate amendment, often gypsum (sometimes sulfur, depending on the chemistry of the soil minerals). Calcium in the gypsum is a stronger cation than sodium, and the calcium cations switch places on cation exchange sites (chapter 4), pushing sodium cations into the soil solution. Once in the soil solution, the sodium cations leach from the soil. But it is much easier to manage a soil to prevent sodification than to reclaim the soil after the damage is done.

HUMAN ACTIVITIES AFFECTING SOIL DEGRADATION

Human activities such as deforestation, intensive tillage, excessive livestock grazing, poorly managed irrigation and chemical fertilizer inputs, and low returns of crop residues accelerate erosion, acidification, desertification, and salinization beyond natural rates on a global scale. Population growth requires more living space and services, so every year natural and agricultural ecosystems are destroyed to build housing subdivisions, water treatment facilities, roads, schools, shopping malls, factories, and recreational facilities. All of these activities contribute to the worldwide trend of soil degradation.

DEFORESTATION

People have always cut down trees. Some trees were harvested to provide wood for fuel and lumber to build houses and ships. Other trees are harvested to clear land for agriculture. Wood is a renewable, carbon-neutral energy source when managed correctly, but when done improperly timber harvesting often leads to soil degradation and other environmental issues.

The effects of tree loss on soil are significant. Trees and shrubs shield the ground from the force of raindrops and provide shade that reduces surface soil temperature, which in turn reduces evaporation. Logging and small-scale removal of trees exposes soil to rain splash. This loosens and dislodges soil particles, eroding soil and creating a more impermeable bare surface, which increases runoff (figure 6–29). On steeper slopes, the topsoil is removed or degraded, and the biological health and productivity of the soil also decreases. Bare soil is exposed to the atmosphere, leading to a rise in soil temperature, which decreases organic matter through faster decomposition and increases evaporation.

Deforestation, as defined by the United Nations, is the permanent removal of trees until less than 10% of the forested land remains. Since the beginning of the Industrial Revolution, humans have removed more than half of the original forest cover on Earth. Each year, an area the size of Greece is deforested (figure 6–30).

In the United States, one early explorer of the Midwest noted a squirrel could jump from tree to tree all the way from Pennsylvania to the Mississippi River without ever touching the ground. Today, a squirrel wouldn't get very far on that journey. Large portions of forests in Ohio, Indiana, and Illinois were cut to provide fuel and lumber, but also to clear more land for producing food for the growing populace. That same kind of deforestation is occurring today in tropical rainforests, which are being cleared to develop cropland to feed a burgeoning population. The challenge is that soils in the tropical rainforest are more highly weathered and have less inherent fertility than those in a temperate climate such as the Midwest.

We also have a greater understanding of the importance of these ecosystems today than in the past. Some of the deforested soils in the tropics are fragile and have low fertility potential. They require a high degree of management and proper inputs to grow crops. If improperly managed, irreversible soil changes occur, potentially preventing crop production. Such lands likely would not recover to their initial forested state, and might become savannas, grasslands, or maybe eroded landscapes with little vegetation. Annually about 13 million hectares of world forests are harvested and converted to other uses, or destroyed by natural events. (See chapter 8 for some current initiatives to limit deforestation.)

It is possible to produce food without destroying forests. *Agroforestry* is a sustainable agricultural practice that integrates trees into crop and pasture systems using relatively low inputs. It is used widely in Africa and Asia. *Silvopastoralism* combines and manages livestock, forage crops, and tree production in one integrated pasture system. Agricultural or horticultural crops can also be grown between rows of woody plants. Such *alley cropping* is increasingly common in tropical countries as a way to diversify crops and help maintain soil. In the United States, this practice is used in the Southeast where cattle are raised under pine trees.

URBAN AND RURAL ISSUES

A farm on the outskirts of a city is sold, and a strip mall takes its place. Every year such urbanization removes more land from food production, putting more pressure on the remaining farmland to produce more crops and livestock. Worldwide, more than 80% of food is produced within 20 miles of where it is consumed, although this is not usually the case in developed countries.

Most food in the United States is imported from all over the country and the world. Most of the grain used in bread, pasta, and similar products is produced in the Midwest, Great Plains, and Pacific Northwest. Rice is produced along the Gulf Coast and in California. Vegetables and orchard crops come from all around the country. Citrus fruit is produced in Florida, south Texas, and southern California. Pineapples and bananas are imported from the Caribbean and South America. Every meal could be a geography lesson.

In developing countries, urbanization is an increasing threat to food production and security. People live near water, especially rivers and lakes. The best soils for producing food are often the same soils where cities are built. As these cities expand, more highly productive soil is converted for urban purposes. Food production is driven farther away from the water, often into soils with more marginal productive capacity, and which may be more susceptible to erosion and other forms of degradation. ▷━━

Figure 6–29. (a) A deforested landscape. (b) Erosion occurring along a field nearby.

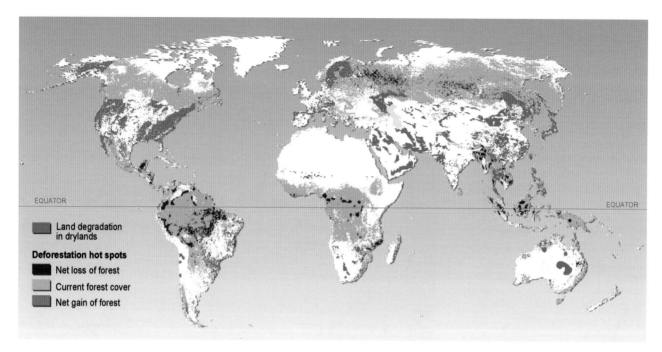

Figure 6–30. While there are significant areas of reforestation, shown as the dark green areas, much of the forest loss is occuring in the tropical areas, with poor quality soils that are prone to erosion.

Urbanization also contributes to runoff and erosion potential. While native landscapes absorb as much as 70% to 90% of precipitation, urban landscapes only absorb about 20%, because of the preponderance of houses, roads, parking lots, and other impermeable surfaces. When native landscapes are converted to urban uses, runoff increases several fold. In addition, the runoff is concentrated, leading to several potential problems such as erosion (figure 6–31), greater frequency and intensity of flooding, and degraded surface water quality.

The lack of green space in urban areas also increases the "heat island" effect: the relatively few plants in many business districts means the areas do not cool down from plant transpiration (see chapter 2) as is the case in native landscapes. As a response, many major urban centers are removing old parking lots, roads, and unused buildings and planting trees and grass in their place. These green spaces have aesthetic appeal, cool down urban centers, and also provide places for water to infiltrate the soil, thereby decreasing runoff and erosion.

Although soil loss due to construction activity is a small fraction of the whole, it is nevertheless a common occurrence. Construction sites are often stripped of all vegetation, leaving bare, compacted soils (figure 6–32) with little infiltration, increasing runoff and erosion potential. The eroded sediments often go directly into streams, rivers, or reservoirs.

Toward Sustainable Agriculture in the United States

In the United States, many government programs promote sustainable agriculture. In 1985, Congress enacted the Conservation Reserve Program, which pays farmers and landowners on a yearly basis for every acre of highly erodible soil that they take out of agricultural production and put into a conservation reserve, by planting either perennial grasses or trees. Also in 1985, the Food Security Act required farmers to develop and implement soil conservation programs to remain eligible for price support and benefits from the federal government. In 1988, the U.S. Department of Agriculture established the Sustainable Agriculture Research and Education program, which provides funding for investigating ways to accomplish sustainable agriculture. As part of the Federal Agricultural Improvement and Reform Act of 1996, the Wetlands Reserve Program pays farmers to set aside land for wetland restoration. The Wildlife Habitat Incentives Program and Environmental Quality Incentives Program were added in 2002 as part of the Food Security and Rural Investment Act. These programs encourage conservation-minded landowners to set aside land for conservation or address pollution problems on their land.

Figure 6–31. Water erosion has washed out part of this parking lot.

Figure 6–32. A construction site presents some serious soil management issues, with the constant wear-and-tear leading to bare, compacted soils that are vulnerable to erosion.

Regulations and construction codes now require mitigation and erosion control practices to be used during construction. *Silt fences* are used to limit how much sediment leaves the disturbed site onto roadways and lawns where it may be tracked into the community or washed into storm drains, rivers, and lakes. However, the fences can be inadequate if used as the only control measure (figure 6–33). The use of *detention basins* is common, especially where extensive construction is taking place and the soil has been stripped of its protective cover and much erosion occurs. Examples are found in construction of housing developments, road interchanges, shopping malls, and office complexes. As the name implies, a detention basin acts as a catchment for debris, sediment, and muddy runoff that would otherwise flow into waterways. *Erosion blankets* are used to cover slopes (figure 6–34), and may be used when the slope is reseeded. A good way to reduce sediment transported off-site involves using several practices (figure 6–35), such as combining sand bags, a detention basin, and a silt fence for more effective control. These and other practices are designed to protect surface water quality and are required by federal regulations under the Clean Water Act.

LAND APPLICATION OF MANURES AND TREATED WASTES

Two of the traditional methods of managing wastes are "out of sight, out of mind," and "the solution to pollution is dilution." While the latter primarily deals with water, the former is a soil issue. Historically, the most common way to dispose of solid wastes was to bury them, while liquid wastes were released into waterways. One of the greatest challenges associated with human, livestock, and food production wastes and byproducts is the quantity of nutrients they contain. Although nutrients may also enter water from fertilizers used in agricultural production (chapter 4), we focus here on the issue of nutrients from wastes.

Manures and byproducts of waste treatment systems are organic, noncommercial nutrient sources that can be used on farms, lawns, and golf courses. These wastes, along with treated wastewater, all contain nutrients that when applied to land may enter surface waters through runoff and erosion or through leaching. Nutrients often found in surface and groundwater include nitrogen, phosphorus, potassium, sulfur, iron, carbon, manganese, boron, and cobalt. When nutrients (especially phosphorus and nitrogen) get into surface waters, nutrient overload (eutrophication) results. You can see evidence of this in the green films of algal blooms (figure 6–36). Appropriate BMPs to control erosion can limit this overload of nutrients in surface waters.

Most meat animals (cattle, swine, poultry) today are fed in confined animal feeding operations (CAFOs), of 500 to more than 50,000 animals. Feed (along with nutrients contained in that feed) is imported from elsewhere. Many dairies and egg-laying facilities also qualify as CAFOs.

Concentrating so many animals in one place greatly increases the efficiency of production. However, the amount of manure and *effluent* (liquid wastes from runoff or washing production floors) produced per area increases several fold with CAFOs. These operations are considered *point sources* of pollution and are regulated as such under federal law. They are required to obtain permits that address how they will manage nutrients in waste to protect surface water and groundwater.

Figure 6–33. A silt fence as the only erosion control method may fail. Silt fences should be used in conjunction with other management tools to limit erosion in construction projects. Can you identify evidence in the photo that this fence is beginning to fail?

The wastes contain essential plant nutrients and are typically disposed of through land application. However, when the wastes are applied improperly or overapplied, the result can be increased nutrient loads in leachates to groundwater and runoff that reaches surface waters.

Further, when wastes are applied to the same land over many years, nutrients may increase to levels that limit plant growth, and increase the potential for leaching into groundwater. Nitrates released from mineralization (chapters 3 and 4) of manure are mobile and can quickly leach, denitrify, or be lost in runoff, causing environmental harm to both water and air quality.

Phosphorus, on the other hand, is less mobile, and so becomes more problematic after many years of manure use, as its concentration builds up to excessive levels over time. When levels are high enough, soil and organic sediments that erode carry enough

Figure 6–34. An erosion blanket has been installed on roadside construction to limit water erosion at this site. Some erosion blankets are used in conjunction with reseeding.

Figure 6–35. Sand bags used in conjunction with a detention basin and silt fence to decrease erosion during a road construction project.

phosphorus to cause algal blooms and fish kills in freshwater systems. Further, the chemical processes that help phosphorus attach to soil particles can be "maxed out," and when that happens, phosphorus leaches into groundwater. For this reason, regulations at the national level and in many states now specifically address phosphorus management in land application of manure and effluent from CAFOs.

Waste management is a suite of practices that includes the land application and treatment by soil of animal, human (residential or municipal), and food-processing wastes in an environmentally sound manner. This BMP considers the waste as a resource and applies the materials at a rate that allows plants to utilize the nutrients and the soil and its microbial community to do the final treatment (remove *pathogens*) and dispersal of liquid. These materials can improve soil quality by increasing tilth, aggregation, water-holding capacity, and beneficial microbial activity by adding organic matter to the soil. Excessive nutrient loading can be avoided if nutrients are applied at the agronomic rate (the nutrient level required by the crop, based on expected yield), adjusted for the potential for phosphorus loading.

Municipal, industrial, and CAFO wastewater treatment systems begin with liquid–solid separation. The "solids" are referred to as *sludge* or *biosolids* and can have as little as 0.5% solids. Slurries are typically 2% to 8% solids, and cake is typically >20% solids. Sludges are dried and sometimes treated further to form biosolids. Sludges, biosolids, and other non-traditional materials increasingly have been used as soil amendments and no- or low-cost fertilizer sources.

Wastewater treatment systems in municipalities and urban areas are centralized—that is, the waste from many homes, businesses, and industries are collected and channeled to the same treatment facility. For example, Albuquerque, New Mexico, has the largest wastewater treatment facility in the state. Wastewater from toilets, showers, dishwashers, and industry all goes into the sewer where it flows to a wastewater treatment facility. There the water runs through screens to remove trash and debris. The screened wastewater flows into a sedimentation basin where the solids settle, and the effluent flows into aeration basins that add oxygen to oxidize nitrogen and organic compounds. Next, the effluent flows into another sedimentation basin where the biosolids are removed. After disinfection and chloride removal, the water is either reused for landscape irrigation or discharged into surface water, typically a river.

The solids (sludge or biosolids) from the sedimentation basins are pumped into anaerobic digesters to produce methane gas, which is used to generate heat and about half the electricity needed to run the plant. After being removed from the digesters, the biosolids are dewatered. Some of these biosolids are disposed of

Figure 6–36. Eutrophication in a small livestock pond.

in landfills, some land applied at agronomic rates, and some are taken to a facility where they are composted with lawn trimmings, chipped tree cuttings, and discarded animal bedding. The materials are placed in long rows and mixed and turned several times for three weeks, allowing microbes to compost the material. The temperature increases to more than 60°C (140°F), which pasteurizes the compost, killing unwanted bacteria and viruses. After curing for three weeks, and screening to remove wood chips, the compost is applied to city parks and school lawns and playing fields. Homeowners can also buy the compost for their gardens and lawns.

Centralized treatment facilities make treating wastes properly somewhat easier, although some safety issues still remain. Biosolids can contribute to higher heavy metal concentrations, mostly from industrial processes. Heavy metals do not degrade but accumulate; therefore, land application needs to be limited so that harmful concentrations are not reached. The Clean Water Act requires city and other large wastewater treatment plants to have an industrial pretreatment program to reduce the amount of heavy metals entering the sewers. Biosolids can also contain pathogens, generally from human and animal wastes, which can be degraded through various treatment systems (such as composting, combined high temparature and high pH treatments, or ultraviolet light treatment). But if the treatment is not properly carried out, the pathogens can pose a threat to those who use the soil and water resources.

Liquid (effluent) and biosolids from municipal wastewater treatment and food processing wastes must meet certain provisions of the Clean Water Act before land application. A current area of emerging concern with the land use of biosolids is the transport and spread of synthetic pollutants in plastics and pharmaceuticals that can end up in biosolids. These compounds are not currently regulated, although the U.S. Environmental Protection Agency is working to identify their risks and propose regulations as needed to protect environmental and public health.

Outside of cities, people typically rely on the soil to effectively, efficiently, and safely treat and disperse wastewater. Decentralized wastewater technology and management refers to treatment and dispersal systems ranging from individual on-site treatment (commonly known as septic systems) to small community collection and treatment systems as well as the process involved in siting, installing, operating, and maintaining the systems. These systems allow treated wastewater to reenter the hydrologic cycle close to where the water was removed—in the case of an individual system often less than a few hundred feet (table 6–2; see also figure 1–3b). These are considered *nonpoint sources* of pollution. (In contrast, CAFOs and centralized wastewater systems collect and treat wastewater on a large scale. The treated wastewater is then released to the environment as a *point source* discharge, effectively removing the water from the area where it was used.)

Although some people may feel septic systems are antiquated and environmentally unfriendly, this is not necessarily the case. Septic systems and soil treatment of our waste have been around for millennia. It is even mentioned in the Bible; see Deuteronomy 23:13. As discussed in chapters 2, 3, and 4, the physical, biological, and chemical properties of the soil combine to make it a great medium for treating wastewater, which we discuss in more detail.

As wastewater flows out of a septic tank into the drain field and soil beneath it, most of the treatment takes place. The septic tank effluent contains pathogens, organic material, and nutrients such as nitrogen in ammonium (NH_4^+) as it enters the drain field. Within the drain field the effluent flows out of drainpipe holes and slowly trickles through gravel. Once it flows into the soil many of the pathogenic bacteria are filtered out or die off. Smaller organisms such as viruses are adsorbed onto soil particles until they too are destroyed. Nutrients such as phosphorus and some forms of nitrogen may be adsorbed in the soil, thus reducing nutrient additions to the environment.

In a properly functioning drain field the soil beneath the trenches is aerobic (contains oxygen). The aerobic conditions are needed to ensure proper treatment of the pathogenic bacteria found in the effluent. The form of nitrogen found in the effluent is ammonium (NH_4^+) (see chapter 3). In the aerobic soil under the drain field the NH_4^+ is converted to nitrate (NO_3^-). Unlike NH_4^+, which is often adsorbed onto soil particles, NO_3^- may flow through the soil and eventually reach the groundwater. A properly functioning septic system will effectively eliminate pathogenic organisms and prevent them from getting to the groundwater, but it will produce NO_3^-.

Some of the NO_3^- that enters the groundwater may be naturally converted to nitrogen gas (N_2) through the process known as denitrification. For this process to occur NH_4^+ is first aerobically converted to NO_3^-. The NO_3^- must move into an anaerobic zone where the bacteria present will convert it to N_2. This conversion only occurs if the correct bacteria are present and there is sufficient food (carbon) for the bacteria to consume. In the natural environment this process can be enhanced in *riparian buffers*, but the actual amount of NO_3^- converted to N_2 is difficult to either predict or quantify.

MINELAND RECLAMATION

There is little doubt that surface and underground mining have dramatic environmental effects on the soil and the surrounding landscape. The materials remaining after coal or ore is removed, known as mine spoils, and the exposed surfaces need to be reclaimed to prevent pollution from the acidic water and metals involved in mining (figure 6–37). Such *reclamation* aims to restore the aesthetic appeal of the landscape and promote plant diversity and the presence of wildlife, in addition to reducing pollution locally and down range from the source. Mine reclamation may require importing fill

Table 6–2. Septic systems (decentralized systems) as a green technology.

Natural disinfection of pathogens (germs). As the effluent flows into the soil most of the pathogenic bacteria are filtered out, adsorbed, or die off. This occurs naturally without the addition of disinfection chemicals such as chlorine.

Reduced nutrient additions to the environment: In a typical household with four people approximately 40 lbs of N and 5 lbs of P will be produced annually; some of this N and P will remain in the septic tank as solids accumulate.

Soil can adsorb or fix nutrients such as phosphorus thus reducing nutrient additions to the environment. Sandy soil and soil with low Fe do not adsorb as much P. As long as no erosion occurs the soil provides a large sink for P from wastewater.

Grasses or other vegetation over the drain field use N and P from effluent as fertilizer. This is evidenced by the green strips commonly seen over the drain field and is another example of reduced nutrient additions to the environment. The remaining N will likely move off site in groundwater as nitrate or it may naturally be denitrified in the soil.

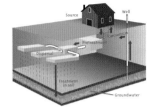

Water recycling: This is not done directly (although some advanced systems are capable of doing so) but indirectly. Consider a lot with house, well, and septic system (recall figure 1–3b). Water enters the house from the well, is used by the residents, and is then flushed into the septic system. After solids are separated out in the septic tank, the wastewater is slowly dispersed into the soil where it is naturally treated. This treated water eventually percolates to the groundwater where it is once again available for use within the house.

Groundwater aquifer recharge: Septic systems disperse the water used in the house into the same watershed. Centralized sewer and collection systems may transport or dispose of wastewater into a different watershed or discharge it to surface waters. This practice could deplete the local aquifer.

Best use of the soil and site: Systems are designed both by practice and by regulation to make the best use of soil and site conditions. This keeps costs and environmental effects to a minimum. It also prevents dense development, keeping more land open and permeable to reduce stormwater runoff.

---

material, topsoil, and organic-rich material to promote plant growth and establishment. In recent times, biosolids have found acceptance in different parts of the country as soil amendments for mineland reclamations because they are either free or cheap to obtain, and they provide a use for the sludge and biosolids municipal wastewater treatment facilities generate.

Leaf compost and other process waste materials such as limekiln dust, fly ash, and paper mulch have also been used in reclamation. These amendments, singularly or in combination, provide nutrients to improve the fertility of the landscape for vegetation, alkalinity to raise the soil pH, and a source of organic material to adsorb toxic heavy metals, making them less available to plants and animals, and to improve water retention to sustain plant growth. New mine permits require assessment of potential topsoil and fill materials onsite as part of the mine's environmental impact statement.

Mine wastes affect soil's physical, chemical, and biological properties and overall soil quality. Wide areas are exposed to erosion from the nature of mining and

Figure 6–37. Reclamation of the Standard Mine Superfund site in Colorado. The contaminants of concern are primarily heavy metals, with samples showing elevated levels of manganese, lead, zinc, cadmium, and copper. The reclamation activities involved efforts to avoid surface and groundwater contamination, revegetation, land use controls to minimize human exposure to contaminants, and monitoring.

construction activities. A variety of technologies aims to minimize the erosion effects. Silt fencing is used to trap sediment and limit how far it is transported. Straw bale barriers are used where temporary berms or diversions are needed. Such barriers act as a sediment filter by allowing water to pass through. Sedimentation (detention and retention) ponds are created to catch runoff from the site. This allows sediment in the runoff to settle and remain on the site. Vegetation on the top and surroundings of the pond can allow sediment-free water to decant off the pond. The ponds should be large enough to catch sediment from major storm events. The water collected may need to be treated or prevented from flowing into nearby surface waters where it could cause damage.

Pyrite (FeS_2) is a common mineral in coal and hard rock mines, and in the western United States, some forms contain arsenic that is a threat to both surface and groundwater quality. The best solution for mine spoils containing pyrite is to bury them, to keep the pyrite from being exposed to the atmosphere where it can be oxidized to produce sulfuric acid. The surface material can then be replanted. Revegetation of disturbed areas, whether temporarily or permanently, can reduce sediment loads and slow runoff from a mine site.

SUMMARY

Soil is an essential component of agriculture, and a slowly renewable resource that can be replenished as it is used. Yet soils form very slowly, and the accelerated erosion in human-altered and managed systems occurs more rapidly than the formation processes that replenish soil. The best management practices discussed in this chapter were developed to help conserve soil in ways that are ecologically compatible and that repair past damage when possible.

This chapter began with a discussion of sustainability. We now return to that thought. The primary challenge in achieving sustainability lies in preventing unrecoverable losses—in other words, **we need to keep our soil in place!** The secondary challenge lies in renewing soil's productivity and its ability to facilitate ecosystem services. We can do this through ecologically appropriate practices.

Implementing conservation practices and BMPs will contribute to a more sustainable agriculture, as well as more sustainable urban and rural development. Conservation tillage, contour farming, strip cropping, grass buffers, crop rotations, mixed system cropping, alley cropping, cover crops, terracing, agroforestry, silvopastoralism, and similar practices are all designed to protect and conserve the soil. Even in urban areas, planting trees and maintaining grass as ground cover helps conserve soil and provides other benefits.

Placing a value on soil conservation is difficult, as it does not fit into traditional economic cost-benefit analysis. Some conservation efforts increase revenues simply because they increase productivity. Some lead to delayed profit. For example, the establishment tree crops typically takes 5 to 15 years before the benefit is realized. Other conservation practices produce no direct monetary profit to the grower, but are done for aesthetic, protective, and ecological reasons. These may lead to higher tourism and recreation incomes, or better human health on a regional scale. These benefits are the most difficult to quantify.

The success and future of sustainable agriculture actually depend on individual landowners, farmers, and ranchers, who need to become convinced that their efforts will work and are affordable, and that in the long run, their own well-being will improve. They need to see and understand that it is to their own benefit and advantage. This may be a key to the future of sustainability: every human being may need to see that one's individual well-being is linked to sustainable development.

Actions toward sustainability include conservation practices, using renewable rather than irreplaceable resources, reducing pollution, and restoring habitats and biodiversity. The challenge lies in balancing the need to feed the growing population by growing crops and providing raw materials for industry on the one hand, with the need to protect the soil as a valuable resource so that it can continue to support civilization in the long term.

ADDITIONAL READING

Hillel, D. 1994. *The Rivers of Eden*. Oxford University Press.

Larson, G. 1999. *There's a Hair in my Dirt*. Harper Perennial.

Magdoff, F.R., and H. M. van Es. 2009. *Building Soils for Better Crops: Sustainable Soil Management*. Handbook Series Book 10. Sustainable Agric. Research and Education.

Montgomery, D.R. 2007. *Dirt—The Erosion of Civilizations*. University of California Press, Ltd.

Online: http://www.fao.org/docrep/012/i1220e/i1220e00.pdf, http://soilhealth.cals.cornell.edu, http://adapt-n.cals.cornell.edu

CREDITS

6–1, J. Lambert; 6–2, 6–3, 6–4, C. Robinson; 6–5, C. Robinson; 6–6, 6–7, USDA-NRCS, NRI; 6–8, 6–9, 6–10, 6–11, C. Robinson; 6–12, USDA; 6–13, B.A. Stewart; 6–14, 6–15, 6–16, 6–17, C. Robinson; 6–18, T. McCabe, USDA-NRCS; 6–19, 6–20, C. Robinson; 6–21, USDA; 6–22, C. Robinson; 6–23; 6–24, USDA-NRCS; 6–25, 6–26, Clay Robinson; 6–27, J. Lambert; 6–28, C. Robinson; 6–29, B. Moebius-Clune; 6–30, Millenium Ecosystem Assessment; 6–31, C. Robinson; 6–32, W. Adewunmi; 6–33, 6–34, Eco-Guard; 6–35, 6–36, C. Robinson; 6–37, USEPA. Boll Weevil, USDA. Chapter opener image, C. Robinson.

Copyright © 2012. Soil Science Society of America, 5585 Guilford Rd., Madison, WI 53711-5801, USA. *Know Soil, Know Life*. David Lindbo, Deb A. Kozlowski, and Clay Robinson, Editors
doi:10.2136/2012.knowsoil.c6

GLOSSARY

ACIDIFICATION The process of soil becoming more acid as basic cations leach from the soil, leaving acidic cations behind. This is a natural process, but can be accelerated by human activities.

AGROFORESTRY Any type of multiple cropping land use that entails complementary relations between tree and agricultural crops and produces some combination of food, fruit, fodder, fuel, wood, mulches, or other products.

ALLEY CROPPING Growing vegetable, grain and/or forage crops between rows of trees.

ARABLE Land that is capable of sustaining crops, where production is practical and economically feasible.

BEST MANAGEMENT PRACTICE (BMP) Any of a group of practices that conserve soil and water resources.

BIOSOLIDS Sludge (municipal treatment plant solids) that has been further treated, tested, and determined to be safe for land application.

BUFFERS (BUFFER STRIPS) Strips of perennial plants, such as grasses, along borders of fields or water bodies.

CAPILLARITY (OR CAPILLARY ACTION) Combination of the attractive forces between a water molecule and a surface (adhesion) and the attractive forces between water molecules (cohesion).

CARRYING CAPACITY A measure of the ability of the soil to support tractors and other vehicles.

CONSERVATION TILLAGE Any tillage sequence, the object of which is to minimize or reduce loss of soil and water; operationally, a tillage or tillage and planting combination which leaves a 30% or greater cover of crop residue on the surface.

CONTOUR CULTIVATION Performing the tillage operations and planting on the contour within a given tolerance.

COVER CROPS Close-growing crop that provides soil protection, seeding protection, and soil improvement between periods of normal crop production, or between trees in orchards and vines in vineyards. When plowed under and incorporated into the soil, cover crops may be referred to as green manure crops.

CROP ROTATION A planned sequence of crops growing in a regularly recurring succession on the same area of land, as contrasted to continuous culture of one crop or growing a variable sequence of crops.

DESERTIFICATION Land degradation where the land area is becoming increasing dry and inhospitable to plants and animals.

DETENTION BASINS Basins constructed to contain storm water runoff, especially from construction sites, to allow sediments to settle prior to runoff water entering surface waterways such as streams and rivers.

DIVERSION A structure or barrier built to divert part or all of the water of a stream to a different course.

EFFLUENT Liquid portion of wastes removed during the wastewater treatment process.

ENTRAINMENT Particles caught and transported in the flow of water or wind.

EROSION (i) The wearing away of the land surface by rain or irrigation water, wind, ice, or other natural or anthropogenic agents that abrade, detach and remove geologic parent material or soil from one point on the earth's surface and deposit it elsewhere; (ii) the detachment and movement of soil or rock by water, wind, ice, or gravity.

GRASSED WATERWAY A natural or constructed waterway, usually broad and shallow, covered with grasses, used to conduct surface water from or through cropland.

GULLY A channel resulting from erosion and caused by the concentrated but intermittent flow of water usually during and immediately following heavy rains. Deep enough to interfere with, and not to be obliterated by, normal tillage operations.

INTEGRATED PEST MANAGEMENT (IPM) A system that uses a combination of biological, cultural, mechanical, chemical, and regulatory tools to control pests such as weeds and insects.

IRRIGATION The intentional application of water to the soil, usually for the purpose of crop production.

NO-TILL Practices by which a crop is planted directly into the soil with no primary or secondary tillage. The surface residue is left virtually undisturbed except where the seed is planted.

PATHOGEN Any disease-producing agent, especially a virus, bacterium, or other microorganism.

POINT SOURCES [OF POLLUTION] single identifiable source of air or water pollution and that has negligible extent.

RECLAMATION A set of activities practiced on disturbed lands such as those disturbed by mining or oil and gas drilling to limit erosion, maintain air and water quality, avoid surface and groundwater contamination, culminating in revegetation with suitable plant species.

REDUCED TILLAGE Tillage system in which the total number of tillage operations preparatory for seed planting is reduced from that normally used on that particular field or soil.

RESIDUE MANAGEMENT The operation and management of crop land to maintain stubble, stalks, and other crop residue on the surface to prevent wind and water erosion, to conserve water, and to decrease evaporation.

RILL EROSION An erosion process on sloping fields in which numerous and randomly occurring small channels of only several centimeters in depth are formed, occurs mainly on recently cultivated soils.

RIPARIAN BUFFER A vegetated area (type of buffer strip) near a stream, usually forested, which helps shade and partially protect a stream from the impact of adjacent land uses. Riparian buffers have become common conservation practices designed to improving water quality and reducing pollution.

SALINIZATION The process whereby soluble salts accumulate in the soil.

SALTATION A particular type of momentum-dependent transport involving: (i) the rolling, bouncing or jumping action of soil particles 0.1 to 0.5 mm in diameter by wind, usually at a height ?15 cm above the soil surface, for relatively short distances. (ii) The rolling, bouncing or jumping action of mineral grains, gravel, stones, or soil aggregates effected by the energy of flowing water. (iii) The bouncing or jumping movement of material downslope in response to gravity.

SHEET EROSION The removal of a relatively uniform thin layer of soil from the land surface by rainfall and largely unchanneled surface runoff (sheet flow).

SILT FENCE An erosion control practice installed on the perimeter of construction sites to limit movement of eroded soil sediments off the construction site.

SILVOPASTORALISM Practices that combine management of livestock, forage crops and tree production in an integrated pasture system.

SLUDGE Solids remaining after effluent (liquids) is removed during the wastewater treatment process. With additional treatment, sludges may be treated further to become biosolids.

SODIFICATION The process whereby the amount of exchangeable sodium in a soil is increased.

STRIP CROPPING The practice of growing two or more crops in alternating strips along contours, often perpendicular to the prevailing direction of wind or surface water flow.

SURFACE CREEP (i) The rolling of dislodged soil particles 0.5 to 1.0 mm in diameter by wind along the soil surface. (ii) The slow movement of soil and rock debris which is usually not perceptible except through extended observation.

SUSPENSION The containment or support in fluid media (usually air or water) of soil particles or aggregates, allowing their transport in the fluid when it is flowing. In fluids at rest, suspension follows Stoke's Law. In wind this usually refers to particles or aggregates ?0.1 mm diameter through the air, usually at a height of ?15 cm above the soil surface, for relatively long distances.

SUSTAINABILITY Managing soil and crop cultural practices so as not to degrade or impair environmental quality on or off site, and without eventually reducing yield potential as a result of the chosen practice through exhaustion of either on-site resources or non-renewable inputs.

TERRACE (i) A step-like surface, bordering a stream or shoreline, that represents the former position of a flood plain, lake, or sea shore; (ii) A raised, generally horizontal strip of earth and/or rock constructed along a hill on or nearly on a contour to make land suitable for tillage and to prevent accelerated erosion; (iii) An earth embankment constructed across a slope for conducting water from above at a regulated flow to prevent accelerated erosion and to conserve water.

WASTE MANAGEMENT The collection, transport, processing or disposal, dispersal, managing and monitoring of municipal, industrial, food-processing, or animal wastes.

WATER CONTROL STRUCTURES Small, adjustable dams placed in drainage ditches or irrigation canals to manage water levels.

CHAPTER 7

SOILS AND BIOMES

As we learned in chapter 1, *biomes* are large geographic regions with similar environments that have distinctive plant and animal communities. A crucial component of biomes is the soil, which supports both the plant and animal life and is, in part, a product of the same environmental conditions. A biome may contain several *ecosystems*—complex communities of organisms adapted to unique environments. *Abiotic factors* in biomes and ecosystems contribute to the unique environments and include the soil, climate (temperature and precipitation), elevation, latitude, aspect, landscape, and geology (parent materials). This chapter will examine 11 terrestrial biomes and the climate and soils generally associated with them (table 7–1, figures 7–1 and 7–2).

Climate has a great influence on a biome's soils, flora, and fauna. The primary climatic factors that make biomes distinct are temperature (table 7–2, figure 7–3) and precipitation (table 7–3, figure 7–4). Temperate climates have distinct warm and cool seasons, while tropical climates have more uniform temperatures throughout the year. Climatic conditions, soil type, *soil temperature regimes*, and *soil moisture regimes* are primarily responsible for the vegetation of a given biome. This chapter briefly describes the different biomes and their characteristics, and more importantly, focuses on the soils commonly associated with them. Study and compare the temperature and moisture regime maps (figures 7–3 and 7–4) to the biome map (figure 7–1) and to the soil orders map (figure 7–2). What patterns can you identify? This question is discussed in this chapter.

FORESTS

Forests occupy nearly a third of Earth's land surface. Ecologically, they are the most complex and diverse systems. Forests produce more oxygen through photosynthesis than any other biome, while removing large amounts of carbon dioxide from the atmosphere, and helping to offset the carbon emissions of fossil fuels. Their trees are a source of wood for fuel and building materials. Forests can be managed to maintain long-term productivity, but many forests around the world are being cut down without replanting and without use of best management practices (chapter 6) to minimize erosion.

CHAPTER AUTHORS
CLAY ROBINSON
WALE ADEWUNMI
DAVID LINDBO

TROPICAL RAINFOREST BIOME

Tropical rainforests are rich, ecological treasure troves located near the equator (figures 7–1 and 7–5) in hyperthermic temperature regimes (figure 7–3) with udic moisture regimes (figure 7–4). They occur in Central and South America, West and Central Africa, and parts of Southeast Asia, and cover about 6% of the land surface. Hosting as many as 1,000 different species of flora and fauna, tropical rainforests are the most diverse biomes. The dense, lush vegetation is arrayed in layers, with trees as tall as 46 meters (about 150 feet). Some plants grow by perching on tree branches (*epiphytes*), while others, such as *saprophytes*, grow by deriving nutrients from decaying organic materials. Still other plants climb high on trees where they can reach sunlight. These as well as many unique plant species are common in tropical rainforests.

Table 7–1. Major global biomes and their related soils.

Biome	% of land	Characteristics	Dominant soil orders
Forests	31		
Rainforests			
Tropical	6	Greatest biological diversity	Oxisols, Ultisols
Temperate	<0.3	Mid to high latitudes, greatest biomass production	Spodosols, Inceptisols, Histosols, Alfisols, Entisols
Temperate forests	8	Generally >1000 mm precipitation	
Deciduous		Mid-latitudes, trees drop leaves in winter or cool season	Alfisols, Inceptisols
Evergreen		Subtropical latitudes, mix of coniferous and deciduous	Ultisols, Spodosols (where parent material is sandy)
Boreal Forests	17	Coniferous, evergreen forests	
		High latitudes in Northern Hemisphere, south of tundra	Spodosols, Andisols, Inceptisols, Histosols
Grasslands	10		
Temperate		Mid-latitudes, tall, mixed and short grasses	Mollisols, Alfisols
Savanna		Tropical latitudes, mixed grasses with occasional trees	Oxisols, Ultisols, Alfisols
Tundra	9	High latitudes, permafrost present	Gelisols, Histosols, Inceptisols, Entisols
Deserts	33	Generally <250 mm precipitation	
Hot		Generally <35°N or S latitude, <1000 m elevation	Aridisols, Entisols, Inceptisols
Cold		High latitude or high elevation	Aridisols, Entisols
Antarctica	14	Mostly ice, limited tundra along coast	
Shrubland	5	30 to 40°N or S latitude, hot-dry and cool-wet seasons	Alfisols, Mollisols, Inceptisols
Alpine	11	Coniferous forests just below timber line, high elevation	Entisols, Inceptisols, Andisols, Spodosols, Histosols
		Tundra above timber line or no forest vegetation, higher elevation without permafrost	Entisols, Inceptisols, Andisols, Spodosols, Histosols
		Tundra above timber line, higher elevation with permafrost	Gelisols
		Never thaws, generally ice, snow cap, glaciers, bare rock	No soil
Wetlands	<1%	Often depressions in landscapes, shallow water tables (aquic moisture regime)	
Fresh water			Histosols or any order with a high water table (Aquic)
Salt water			Histosols, Entisols, Inceptisols

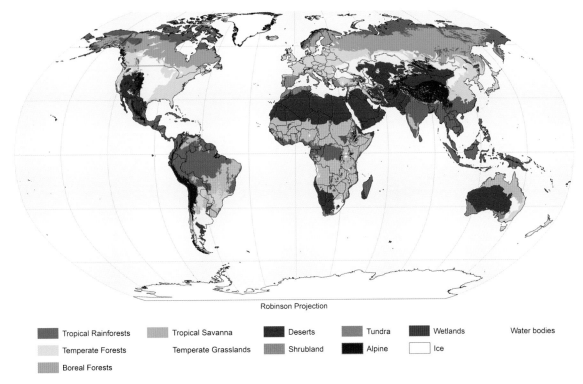

Figure 7–1. World biomes.

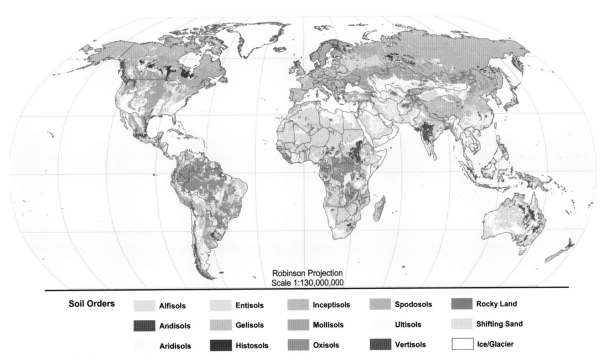

Figure 7–2. World soil orders.

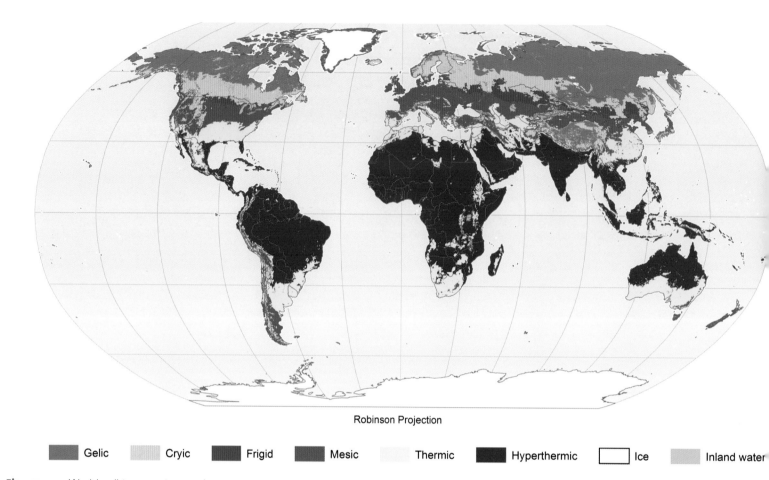

Figure 7-3. World soil temperature regimes.

Table 7-2. Soil temperature regimes.

Temperature regime	Temperature
Gelic	<1°C (<34°F), permafrost common
Cryic	0–8°C (32–47°F), no permafrost
Frigid	0–8°C (32–47°F), summer 6°C warmer than winter
Mesic	8–15°C (47–59°F)
Thermic	15–22°C (59–72°F)
Hyperthermic	>22°C (>72°F)

Table 7-3. Soil moisture regimes.

Moisture regime	Temperature
Aquic	Usually near saturation, essentially no available oxygen
Udic	Soil generally moist, not dry in any part for 90 cumulative days; can grow a summer crop, such as corn, without irrigation
Ustic	Deficient moisture: dry for 90 or more cumulative days, but not dry for more than half of the growing season; cannot grow a summer crop without *fallow* to store soil moisture
Xeric	Deficient moisture: dry summer, wet winter
Aridic	Usually dry, moist less than 90 consecutive days

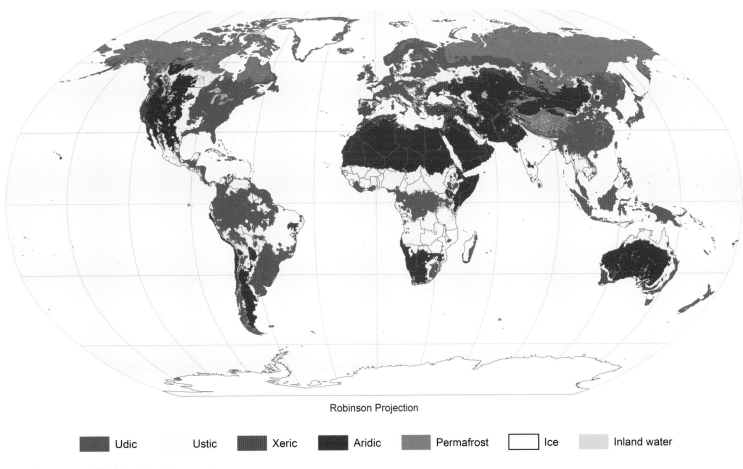

Robinson Projection

■ Udic ■ Ustic ■ Xeric ■ Aridic ■ Permafrost □ Ice ■ Inland water

Figure 7–4. World soil moisture regimes.

Figure 7–5. A rainforest in Costa Rica.

Which Came First, the Soil or the Biome?

Terrestrial biomes cannot form apart from the soil; plants require media in which to grow. The very factors (CLORPT, chapter 2) that form the soil work in tandem with the soil to develop ecosystems and biomes. Soil orders, the highest level of soil classification (see chapter 5), correspond to both regional and global biome definitions.

Most of the rainforest's nutrients are held in the biomass of the lush vegetation. When rainforest flora and fauna die, decomposers rapidly release the nutrients, which are quickly assimilated or *immobilized* into new plant growth. The result is a nutrient-poor soil despite the lush appearance of high fertility. The forest echoes with the noise of colorful insects and birds. Varied animals have adapted to the tropical rainforest in their ability to jump from tree to tree or swing from branches, for example. Some animals, such as fruit bats and toucans, live almost entirely on fruit; in fact, this is the only biome where animals can have a specialized diet based on fruits, because of their availability year round. Bananas, mangos, papayas, pineapples, and avocados are some of the fruits that originated in tropical rainforests.

The prominent soils of the humid tropical rainforest biome are Oxisols and Ultisols (figure 7–2) (see chapter 5). Despite the lush vegetation, Histosols are generally not found in tropical rainforests because the high temperature results in rapid decomposition, so little organic material accumulates. Oxisols (figure 7–6) are highly weathered soils with little visible differentiation in the horizons. Oxisols are typically bright red in color from the presence of iron and aluminum oxides (chapter 2). Intense weathering and leaching remove most of the silica, resulting in a B horizon composed of oxides (Bo) rather than clay. Ultisols also are highly weathered, but not as much as Oxisols. Ultisols (figure 7–7) have an accumulation of clay in subsurface horizons (Bt, chapter 2) and are also rich in iron and aluminum oxides.

The high amounts of iron and aluminum oxides give Oxisols the ability to tightly bind any phosphorus present in the soil. These soils are low in nutrient cations (bases such as calcium, magnesium, potassium) (see chapter 4) and are therefore moderately acidic and have low natural fertility. When these soils are cleared of native vegetation and used for cultivation, productivity declines rapidly in the following years because the soil has no native reserve of nutrients to resupply growing crops every season. Organic matter from plant tissues is the main natural fertility source. If not managed carefully, Oxisols can have severe structural changes that limit water infiltration and root growth. However, fertilization, particularly high amounts of lime ($CaCO_3$) and phosphorus, can increase crop yield and maintain Oxisol productivity.

The Ultisols of this biome have appreciable quantities of clay minerals in the B horizon and more base cations than do the Oxisols. Weathering of the clay minerals and leaching of the base cations result in acid topsoil and subsoil. As with Oxisols,

Figure 7–6. Oxisol. Note the lack of clearly visible horizons below the surface A horizon.

Figure 7–7. Ultisol. The color results from iron oxides coating the individual particles in the soil.

Ultisols are low in plant nutrients, but nutrient management can result in fertile and productive agricultural soils.

Management of tropical rainforest soils requires that vegetation remains on the land to reduce erosion from heavy rainfall and preserve the ecology for future productivity. In this regard, farmers often practice mixed cropping (intercropping) so that when one crop is harvested, another (growing) crop remains. Cultivation of tree and perennial crops and mixed cropping practices can be carried out with equal success on Oxisols and Ultisols.

A form of subsistence agriculture called *shifting cultivation* is common in tropical rainforests (figure 7–8). It involves slashing (cutting) and burning existing vegetation (see chapter 8). Farmers cut trees in one area, and reserve some of the wood for fuel and building materials. The remaining wood and residues are burned, releasing base cations (chapter 4) and increasing the nutrient supply for crops that will be grown. The area will be cropped for a few years until the fertility of the soil is depleted. Then it will be abandoned, allowing vegetation to grow again. Farmers then cut a new area, and the process begins again. After 20 to 25 years, the rotation returns to the original area.

The trees and vegetation bring up nutrients from deep in the soil and return them to the surface through leaf drop and decomposition, and through burning the trees after cutting. In *alley cropping*, or *agroforestry*, farmers leave some of the trees in place and grow food crops between rows of trees. *Deforestation* (chapter 6) is becoming more common as the ever-increasing human population needs an ever-increasing supply of fuel and food.

TEMPERATE RAINFOREST BIOME

Temperate rainforests are not common, now occupying less than 0.3% of Earth's land surface. As much at 90% of these forests have been felled and no longer exist. Remaining temperate rainforests are found in the Pacific Northwest of the United States, Chile, Japan, New Zealand, southern Australia, Norway, and the United Kingdom, generally from about 30 to 55° north or south of the equator in coastal areas with cold ocean currents. Temperate rainforests have much less biodiversity than their tropical counterparts, but produce much more biomass. Common tree varieties include conifers such as western hemlock, western redcedar, Sitka spruce, Douglas fir, and coastal redwoods, some as tall as 100 meters (330 feet). Deciduous trees, such as maples and dogwoods, and shrubs, such as wild currant and huckleberry, grow in the understory. Ferns, lichens, and mosses (epiphytes) hang from tree branches. Lichens, algae, mosses, oxalis, and grasses form a rich carpet on the surface soil.

The soils are typically Andisols, Spodosols (see boreal forests), and Alfisols (figures 7–1 and 7–2) depending on parent material. Those parent materials include volcanic ash, loess, residuum, glacial, and fluvial sediments (see chapter 2).

Figure 7–8. Slash and burn agriculture in Brazil. The tropical forest has been burned to plant corn.

TEMPERATE FOREST BIOME

Temperate forests are found between about 23 and 50° latitude (table 7–1). Temperate climates receiving more than 100 centimeters (40 inches) of annual precipitation generally support temperate forests, depending on the elevation and latitude. The *frigid* and northern extents of *mesic* temperature regimes and alpine regions support evergreen (coniferous) forests, which will be discussed later. These coniferous forests are important sources of wood for lumber and the pulp industry, accounting for about 40% of all softwood timber harvested in the United States. Deciduous forests are characterized by a dormant season in which the trees drop their leaves. They are dominated by tree varieties such as maple, oak, alder, ash, birch, beech, elm, chestnut, hickory, walnut, and pecan. Deciduous forests primarily occur on glacial till, loess, eolian, residuum, and alluvium parent materials.

Northern deciduous forests have *mesic* (table 7–2) temperature regimes (figure 7–3) and *udic* to *aquic* (table 7–3) moisture

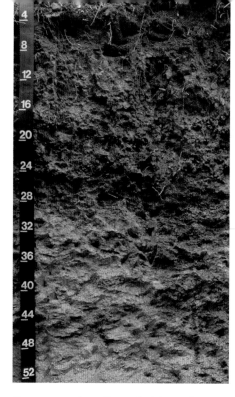

Figure 7–9. A northern deciduous forest Alfisol with a thin dark A horizon above an E horizon (almost white horizon) over a Bt horizon with illuvial clay accumulation.

Figure 7–10. A deciduous forest Inceptisol with a thin A horizon (0–10 centimeters, 0–4 inches) over a Bw horizon with slight development.

regimes (figure 7–4). Southern forests are warmer and have *thermic* (table 7–2) temperature regimes (figure 7–3). Southern forests occur as deciduous, coniferous, broadleaf evergreen, or mixed forests. Mats of decomposing leaves in forests often result in a shallow organic surface horizon. The decomposing leaves release acids, which make minerals and nutrients more soluble. As water moves through the soil profile, an E horizon (eluvial) may form as minerals and nutrients are eluviated, or moved out of the horizon. The landscape position affects the amount of development in the B horizon.

Northern deciduous forests commonly have Alfisols and Inceptisols (table 7–1, figures 7–2, 7–9, and 7–10). Alfisols occur on stable landscapes and have more development than Inceptisols, which occur on steeper slopes and younger landscapes. Alfisols are similar to Ultisols, with a subsurface accumulation (of illuvial) of clay (Bt, chapter 2), but are less weathered and have a greater base saturation. Ultisols in tropical rainforests are less likely to have E horizons. Inceptisols have some development in the B horizon (Bw, chapter 2).

Alfisols are more common in northern forests than southern forests because cooler temperatures result in less weathering. Southern forests typically have Ultisols

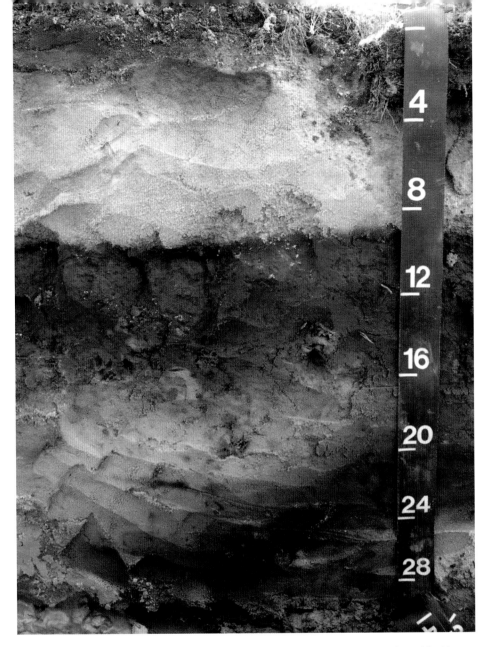

Figure 7–11. A wet Spodosol from a long leaf pine forest. Notice the A horizon (0–10 centimeters, 0–4 inches) over an E horizon (10–23 centimeters, 4–9 inches) over a spodic horizon (23–40 centimenters, 9–16 inches).

(figure 7–7), or Spodosols (figure 7–11) if the parent material is sandy in texture.

The pH of Alfisols typically ranges from about 5.8 to 6.8, while Ultisols are more acidic, ranging from 5.0 to 6.5. Alfisols may be productive farmlands if the soil is supplemented with fertilizers to preserve productivity, since precipitation beyond plant requirements may move through the root zone and leach nutrients out of the soil. Ultisols require more intensive management to be productive farmlands because they have lower pH and lower natural fertility. Spodosols are sandy, with low pH (<5.0) and low levels of nutrients.

Many U.S. temperate forests were logged in the 1800s as the Midwest was settled. These soils, mainly Alfisols, are now productive farmlands but often require additions of limestone to improve nutrient availability by raising the pH to 6.5 or 6.9, depending upon crop requirements (see chapter 4). The soils typically have low organic matter levels, and fertilizer applications are required to supply other nutrients. Landscapes cleared for farming typically have 2% to 15% slopes, and farmers must use best management practices (BMPs) to control water erosion on these soils (chapter 6). Reduced and no-tillage systems, strip cropping, crop rotation, cover crops, and grassed waterways are commonly practiced on these Alfisols and Inceptisols. Some of the lands on steeper slopes are being reforested today.

In the southeastern United States, most of the cotton, sugarcane, and tobacco plantations that were cleared with slave labor occur on Ultisols. All the BMPs used on Alfisols also must be used on Ultisols. Some of the more eroded and degraded lands on steeper slopes are being reforested.

CONIFEROUS/BOREAL FOREST BIOME (TAIGA)

Boreal forests cover about 17% of Earth's land surface (table 7–1, figure 7–1). These forests occur south of the tundra located in gelic and cryic temperature regimes of the northern hemisphere from about 58° north in eastern Canada to 68° north in Alaska (table 7–1). In Europe and Asia, the northern boundary is along the Arctic coastlines except in Siberia, where it cuts deeper inland. Boreal forests are also known as *taiga* (the Russian word for forest). These are evergreen forests dominated by conifers such as fir, spruce, pine, and larch, which occur as swaths of uniform species over wide areas. The forests occur on glacial till, lacustrine, alluvial, eolian, residuum, and volcanic parent materials.

Spodosols, found on sandy soils, are the predominant soil order in boreal forests (figures 7–2 and 7–11). Where the parent material is not sandy, Inceptisols and Entisols are common (figures 7–10 and 7–12). On plains, moraines, or outwash surfaces, the soil parent materials are primarily glacial tills or alluvial materials derived from till.

Needles from the coniferous trees produce organic acids as they decompose, contributing to the formation of the biome's Spodosols. Nicknamed "white earths," Spodosols are characterized by an ash-colored layer (E horizon) below the O or A horizon. The unique ash color is due to the eluviation of organic matter as well as iron and aluminum compounds from the E horizon, as strong organic acids dissolve the compounds, and percolating water carries them out of the horizon. The translocated or eluvial iron and organic compounds from the E horizon accumulate (illuviate) in the B horizon, creating a spodic horizon (Bh, Bs, or Bhs, chapter 2) with reddish black to reddish brown colors. These soils are acidic, sandy, and low in fertility because the decomposing conifer needles are low in base cations (chapter 4), and the acids produced during decomposition have leached many of the mineral nutrients.

GRASSLANDS

Grasslands cover about 10% of Earth's land surface. They come in a variety of types, and are called by a variety of names, depending on where they are located: steppes, prairies, pampas, veldt. Grasslands are responsible for producing most of the food people consume, in grains such as corn, wheat, and rice, and in meat animals such as cattle, goats, and sheep. Grasslands can be managed to sustain productivity with BMPs for erosion control when they have been plowed for crop production, or with carefully designed grazing management programs in native pastures.

Savannas are also dominated by grasses, but contain scattered trees and occur in many regions. Small areas of savannas in subtropical regions are transition zones between forests and grasslands, while tropical savannas are a distinct biome and cover vast regions.

TROPICAL SAVANNA BIOME

Tropical savannas feature tall grasses interspersed with open canopy shrubs, palms, and trees. They are able to withstand several months of dryness or drought following the rainy season. Savannas cover large areas of the Sahel in Africa, with smaller areas on other continents (table 7–1, figure 7–1). In East Africa this biome is home to many large and small grazing animals, as well as predators and scavengers, such as lions and hyenas. These animals thrive because of the abundance of

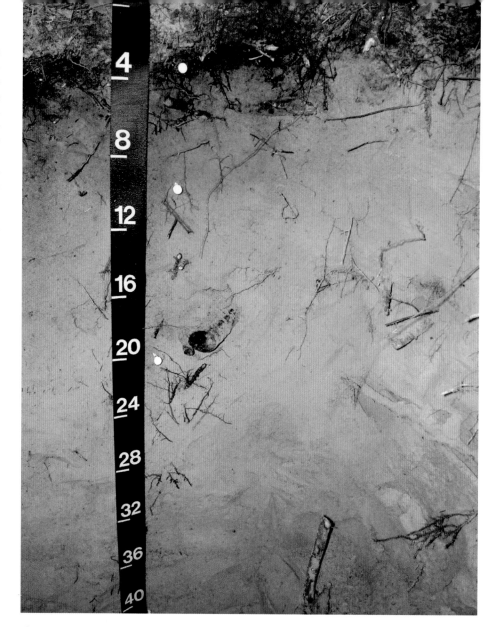

Figure 7-12. A forested Entisol with no soil development below 25 centimeters (10 inches).

vegetation for the herbivores, which are, in turn, available as prey for the carnivores. Termites are abundant in tropical savannas, and so are their conspicuous mounds (*termitaria*) across the landscape. These insects play a huge role in moving soil materials.

Savannas are typically found on old landscapes, which are often highly weathered. Alfisols and Ultisols (with some Oxisols) occur in this biome; Alfisols are most common (figure 7–2). In the tropics, these soils are low in fertility because of the age of the landscape and the effect of leaching from heavy rainfalls during rainy seasons. The warm temperatures and adequate moisture promote decomposition in the rainy season, so these soils are low in organic matter. Laterites, which are hardened, reddish to dark brown, highly weathered soils rich in iron and aluminum oxides, are common in Africa, India, and western Australia. They can be found just below the soil surface; when exposed, they appear as massive rock outcrops, depending on the size and landscape placement. In India, they are cut and used as bricks for building construction.

TEMPERATE GRASSLAND BIOME

Temperate grasslands are among the most productive soils in the world (table 7–1, figure 7–1). The tall- and short-grass prairies of the U.S. Great Plains are examples of this biome (figure 7–13). The dominant grasses in short-grass prairies are less than 50 centimeters (20 inches) tall. They occur in *ustic* (table 7–3) moisture regimes with less precipitation than tall-grass prairies, which occur in udic moisture regimes (figure 7–4), and in which the dominant grasses are taller than 50 centimeters. About 13% of U.S. grasslands are tall-grass and oak savannas that occur in transition zones between forests and prairies. Before settlement, grasslands covered about half of the contiguous 48 states, with 95% of those west of the Mississippi River. About half of these grasslands have been converted to cropland or other uses. The U.S. tall-grass prairies and northern tall-grass savannas are especially rare.

Mollisols are the dominant soil order in this biome (figure 7–2). Most of the Mollisols in the world are found in grassland biomes in temperate regions with mesic and thermic temperature regimes (table 7–2). The primary feature of Mollisols is a dark thick A horizon, due to high amounts of organic matter. The organic matter comes from the grass roots that die and decompose each year (chapter 2). Mollisols do vary. Tall-grass prairie soils are darker, and the dark colors extend deeper into the soil than in the short-grass prairie soils (figure 7–14). The organic matter coats the soil particles, making them appear dark. Receiving less precipitation, short-grass prairies produce less biomass and have less organic matter than tall-grass prairies. Water does not percolate as deep in these soils. Horizons of salts, calcium carbonate, or gypsum often are found in soils of short-grass prairies, while these compounds are typically leached below the soil in tall-grass prairies.

Some soils in the temperate grasslands of the southern Great Plains have sandy or sandy loam textures. Organic matter does not bind as tightly with sand particles as with clay particles. These soils typically are Alfisols (figure 7–15), which cover significant areas in both temperate and tropical grasslands (figure 7–2).

Most of the cereal grains (including rice, wheat, corn, sorghum, millet) and pulses (soybeans, field beans, peas, cowpeas, peanuts) in the world are grown on grassland soils that have been plowed and converted to croplands. These soils generally are naturally fertile and have high levels of organic matter. They provide habitat for a wide variety of grasses and wildflowers, as well as microfauna, earthworms, and wildlife. Before settlement, the grasslands of the Great Plains were home to huge herds of bison, antelope, deer, and elk. These animals wintered in the south or in protected canyons with a year-round water source. When spring rains started, the grazers followed the rains and the growing grass north in summer, then moved south again in fall, eating the grass that had regrown in

Figure 7–13. An expanse of prairie in Kansas.

their absence. The bison were almost exterminated to make room for grazing cattle and sheep to graze, and for homesteaders.

MANAGING GRASSLANDS

Grasslands pose several unique challenges for land managers. The challenges include grazing management, desertification, wind and water erosion when grasslands are converted to cropland, and irrigation management on cropped lands in drier regions.

Many grasslands are used as grazing lands, especially in the short- and mixed-grass prairies and the savanna (figure 7–16). Historically, many of these lands were overgrazed, leading to erosion and land degradation. In many developing countries, land is not privately owned, and no individual or group is responsible for managing grazing. Overgrazing is common, which weakens the grass, so regrowth is not as vigorous. As the amount of vegetation on the surface decreases, the soil is more susceptible to erosion by wind or water. When these conditions combine with drought, vegetative cover continues to decrease, and *desertification* becomes a concern.

Many grasslands in the United States are privately owned. Range management methods have improved greatly in the

Figure 7–14. Mollisols will look different depending on if they are from (a) tall-grass prairie or (b) short-grass prairie. The thick A horizon in each profile forms from the input of grass roots that die and decompose on an annual basis.

Figure 7–15. A grassland Alfisol typically has a sandy texture and lighter colors in the A horizon due to less organic matter.

last 40 years, and many landowners have adopted best management practices. Some common practices include

- identifying the range ecosite (specific combinations of vegetation on certain soils and landscapes),
- determining vegetative biomass production (how many kilograms of grass or forage grow per hectare in a year),
- adjusting stocking rates to match biomass production, and
- rotational grazing, allowing land to rest and grasses to recover between grazing cycles.

Most of the grasslands east of the Rocky Mountains are privately owned, while the majority of western grasslands are owned and managed by the government, which leases the land to range managers and requires them to use the best grazing management practices available.

Low precipitation in many grassland regions (ustic or *xeric* moisture regimes, figure 7–4) limits the production of annual crops. In these areas, irrigation is common when water is available from rivers or aquifers. Irrigation is becoming more common in udic moisture regimes (areas with greater precipitation), a bit like an insurance policy to reduce the risk of yield losses during brief droughts. The quality of the irrigation water has an effect on the sustainability of agricultural production. Irrigation water with high salt or sodium

Figure 7–16. Grasslands are often managed for use as grazing lands.

Fire in Ecosystems

Fire is a natural process in native grassland and forest biomes. During dry seasons or periods of drought, lightning starts fires, which consume dead grasses and forest understory. Periodic fires keep the fuel supply low, so the fires are smaller and cooler. Small fires have little adverse effect on trees, perennial grasses, and soil. These periodic fires actually maintain the health of the vegetative community and limit encroachment of invasive species.

However, fires remove valuable surface cover, leaving soil more susceptible to wind and water erosion. In mountainous areas, when a rainstorm follows a forest fire, rivers turn black with ash and soil. Hot fires and severe burns damage perennial grasses, trees, and shrubs, and may create *hydrophobic soils*. Water cannot easily infiltrate into hydrophobic soils, so there is more runoff, leading to more erosion. After a severe burn in a forest, there is less shade, so the soil surface gets hotter, which changes the types of vegetation that grow back, sometimes even changing a mountain forest into a meadow. Revegetation and erosion control are important best management practices following hot fires and severe burns.

People interfere with these natural processes by putting out fires to protect their homes, businesses, and towns. Putting out fires allows more fuel to accumulate, especially in forests. When fires do occur, they are more severe and cause more damage. Suppressing fires may also change vegetative succession. In the some parts of the southern Great Plains, eastern redcedar is encroaching on mixed-grass prairies where fire has been suppressed. In various places, mesquite, prickly pear, sagebrush, juniper, and other species alter ecosystem value and function due to fire suppression.

The grazing and economic value of thousands of hectares of rangeland has decreased, the aesthetic value of wilderness areas has been diminished, and the intensity of forest fires has increased as a result of decades of fire suppression policies.

Forest and range managers now understand the value of fire in ecosystem management. Controlled burns are used to reduce understory vegetation and fuel supply, decrease encroachment of invasive species, and increase the grazing value of land for cattle and sheep. Climatic conditions are crucial to success; if the humidity is too low or the wind too high, a controlled burn quickly can turn into a wildfire.

Every year, wildfires affect thousands of hectares of grasslands and forests. Wildfires destroy homes, threaten towns, cause accidents, and affect air quality, sometimes for hundreds of kilometers. Severe wildfires may race across the landscape in excess of 80 kilometers (50 miles) per hour and claim human and animal lives. Too many of these fires are started by carelessness: throwing a cigarette out a car window, using fireworks, burning trash, or even having a cookout on a windy day. Do your part to prevent wildfires, and leave controlled burns to professionals.

On the rim of the far side of this canyon (upper photo), there were almost no shrubs in 1980. The shrubs are readily evident in the second photo, taken in 2004, after fire suppression efforts have allowed encroachment of junipers and mesquite.

content will increase the concentration of salts and sodium in the soil. Salts are detrimental to plant growth, while sodium is detrimental to soil properties (chapter 6). Appropriate irrigation management is necessary to sustain agricultural production without compromising the soil.

TUNDRA

Tundra biomes are characterized by cold climate, which occur either at high latitudes in the northern hemisphere or at high elevations in alpine regions atop mountain peaks. Moisture in this biome group is in the form of fog, snow, or ice (frozen water). Tundra is the simplest biome in terms of species composition and food chain. Tundra comes from *tunturia*, a Finnish word for barren lands.

ARCTIC BIOME

Arctic tundra occupies about 9% of Earth's land surface (table 7–1), and is the northernmost of the biomes (figure 7–1), located north of boreal forests, mainly in the northern latitudes surrounding the Arctic Ocean. The arctic tundra is a treeless environment, characterized by *permafrost*, strong drying winds, a very short growing season (of less than 50 days per year) and very shallow soils. Trees cannot penetrate the frozen soil, but low and ground-hugging plant species are common. The lack of tall trees means winds are unimpeded, strong, and desiccating. The sparse vegetation supports a few large and small mammals, such as caribou, muskoxen, and showshoe hares. The small mammals hibernate in burrows in the winter, while the larger grazers and predators such as wolves, wolverines, and polar bears migrate south to the boreal forests. The vast majority of tundra occurs in upland areas, where the slopes provide drainage and the depth to permafrost is greatest.

Soils of the arctic tundra region are young and have little horizon differentiation. The common order, Gelisols (figures

Figure 7–17. Alaskan tundra, showing patterned ground, which forms from the annual freeze–thaw cycles in the upper portion of the soil and permafrost.

7–17 and 7–18), is characterized by the presence of permafrost within 1 meter (40 inches) of the soil surface (or within 2 meters if frost churning is present). Because of the prevailing cold temperature regime, decomposition is very slow and much of the organic material in the soil is not fully decomposed. Histosols occur in the tundra biome where the soils do not contain permafrost and have thick accumulations of organic matter. Gelisols store large quantities of organic matter, which has implications for global warming. As temperatures rise and permafrost melts, the organic matter will likely decompose, releasing additional carbon dioxide to the atmosphere. Gelisols will move north as global temperatures rise.

Gelisols present unique challenges to building and construction (figure 7–19). Freezing and thawing above the permafrost creates unstable soil. Once disturbed, Gelisols become even more unstable as

Figure 7–18. Gelisols are permafrost soils. Organic matter decomposition is slow due to cold temperatures.

non-uniform melting of permafrost may occur. This is the reason the trans-Alaska oil pipeline is built above ground on piers rather than buried below ground. Agriculture is uncommon in this biome because of the short growing season and cold soils.

ALPINE BIOME

Alpine regions are found in high mountain ranges on every continent but Australia. They share common elements with ice caps, arctic tundra, and boreal forests (figure 7–1, table 7–1). Just as temperature cools with increasing latitude, it also cools with increasing elevation, about 10°C for every 1,000 meters (18°F for every 3,300 feet). It might be 35°C (95°F) in Colorado Springs, Colorado (1,840 meters or 6,035 feet), on a warm summer day, while at the top of Pike's Peak (4,302 meters or 14,110 feet), it could be 10°C (50°F). There are also differences between alpine and arctic regions. The freezing and thawing cycle is *diurnal* and not seasonal as in arctic tundra. The length of daylight and seasons surrounding alpine tundra are different as well.

Alpine biomes have the most diverse climatic conditions of all the biomes. Several temperature and moisture regime combinations may be present on the same mountain, depending on elevation and *aspect* (direction the slope faces). Precipitation depends on the prevailing wind direction and slope; a lot of rainfall occurs at intermediate elevations while fog and snow are more common above the tree line. Latitude makes a big difference as well (table 7–4).

Figure 7–19. Gelisol soils freeze and thaw, making for unstable foundations for houses.

The diverse climatic conditions result in diverse vegetative communities. Shrublands may be present at the base of the mountain, changing to coniferous forests with occasional grassy meadows as the elevation increases. The conifer tree species change, then give way to alpine tundra with tussock-like grasses above the tree line, which may give way to ice. Only warm-blooded animals, such as marmots, elk, and mountain goats, and some insects live in alpine tundra in the summer, migrating into the forest or hibernating in the cold season. Such animals have adaptations to withstand the cold and low oxygen levels.

Table 7–4. Characteristics of five mountains in the western hemisphere.

Mountain	Latitude	Elevation	Elevation of occurrence			Parent material source	Common soils
			Lowest glacier	Alpine tundra	Coniferous forests		
			m				
Chimborazo, Ecuador	1°S	6310	4600	Above 4600	3600 to 4600	Volcanic ash	Andisols, Spodosols
Pico de Orizaba, Mexico	19°N	5636	4700	Above 4300	3200 to 4300	Volcanic ash	Andisols, Spodosols
Pike's Peak, Colorado	38°N	4302	None	Above 3500	1900 to 3500	Granite	Entisols, Inceptisols
Mount Rainier, Washington	46°N	4392	2100	Above 2100	Below 1800	Volcanic ash	Andisols, Spodosols
Denali, Alaska (Mt. McKinley)	63°N	6194	300	1200 to 1500	Taiga below 1200	Granite	Entisols, Inceptisols, Gelisols, Histosols

Soils in this biome generally are weakly developed because of the low temperatures. Spodosols and Andisols (figure 7–20) are the exception to this general statement. Under alpine conditions, the weatherable volcanic parent materials associated with Andisols and the organic acids associated with Spodosols lead to soils with more highly developed horizons. One rather interesting soil that may occur in this biome is a type of Histosol, called Folist, which is not saturated for long periods of the year. Folists form as organic matter accumulates because decomposition is slower in the cold environment.

Alpine biomes are fragile due to the extremes in climate. Since they are also not suitable for most human activities, management is not as critical. However, overuse by vehicles and foot traffic can cause severe erosion once the protective vegetative surface is damaged or removed (figure 7–21).

DESERTS

Deserts are defined primarily by the amount of rainfall, or the lack of it, they receive. Obviously the driest of all biomes, deserts are also diverse (figure 7–22). They may be hot and dry (the Sahara) or cold and dry (Antarctica). They may be sandy or rocky, on coasts or inland. Deserts are not devoid of life but support plants and animals that have adapted to the harsh conditions, either by tolerating or avoiding the extremes of the desert conditions.

Deserts occur on every continent but Europe, covering about 19% of the six habitable continents, and 33% of Earth's land surface when Antarctica is included (table 7–1, figure 7–1). Most deserts, other than Antarctica, are located between about 10 and 40° latitude north or south of the equator. Many deserts exist in the rain shadows of mountain ranges. In the western United States, deserts occur in the *aridic* moisture regimes near mountain ranges (figures 7–1 and 7–4).

Deserts are inhabited by plants that have adaptive mechanisms to reduce water loss—for example, thickened, waxy, or small leaves. Vegetation varies, but may include species of cacti, small shrubs, and small annual and perennial grasses and broadleaf plants, although the total vegetative surface cover is often less than 20%. Cacti have a unique photosynthetic process that allows them to open their stomata for gas exchange at night when the relative humidity is higher, so they lose less water to transpiration. Many cacti are capable of storing water in their thickened, fleshy leaves, while other cacti are round and do not have true leaves. As in the cartoons, coyotes chase jackrabbits and roadrunners in the desert. Many reptiles, insects, and arachnids call hot deserts home.

The Atacama Desert in Chile is the driest desert in the world. Antarctica is almost as dry (figure 7–22b), covered by more than 1.6 kilometers (1 mile) of ice in places, and receiving less than 20 centimeters (8 inches) of precipitation along the coast and much less inland. As ice and glaciers cover most of the continent, very little soil is present. Some tundra occurs in the coastal regions, which is home to most of the organisms that live there: algae, bacteria, fungi, mites, penguins, and seals. Only two species of vascular plants have been found. Similar to Antarctica, much of Greenland is a desert.

The dry conditions of the desert limit vegetation, and since there is little vegetation on the surface to protect the soil from erosion, many deserts have "stone pavement" (figure 7–23) on the surface. This happens when wind and water remove all the finer particles from the surface and leave behind gravel and rocks that are too big to be moved. The impact of raindrops on unprotected soil can create surface

Figure 7–20. An alpine Andisol. These soils have a low bulk density and high phosphorus fixing capacity.

Figure 7–21. Alpine soils are vulnerable to erosion.

Figure 7-22. Not all deserts look the same. Both (a) the U.S. Southwest and (b) the dry valleys of Antarctica are classified as deserts.

crusts, especially when sodium is present. These crusts have platy structure (figure 7–24) and limit water infiltration.

Despite the limited numbers of vascular plants, some desert soils develop a biological, or cryptogamic, crust (figure 7–25). These living crusts are formed by communities of bacteria, algae, mosses, and lichens. The crusts protect the soil surface from raindrop impact and erosion, and may increase infiltration of rain. Sometimes desert soils get so dry they repel water rather than absorb it, a phenomenon known as *hydrophobicity*.

Even though deserts receive little total annual precipitation, the rain often comes in intense thunderstorms, which may drop 5 to 10 centimeters (2 to 4 inches) of rain or more in one hour. This exceeds the infiltration rate of the soils, creating large amounts of runoff and flash floods. The runoff water must go somewhere and erodes soil as it goes, quickly producing gullies.

Most desert soils are Aridisols (figure 7–2 and 7–26) and have accumulations of clay (Bt), calcium carbonate (Bk), gypsum (By), and/or sodium and clay (Btn). The rest are typically Entisols, with no horizon

Figure 7–24. Salt crust.

development. The soils have little organic matter and light-colored A horizons. Most scientists agree that the B horizons in desert soils formed in previous climates, such as ice ages when rainfall patterns and vegetation were much different than they are today. In the current climate, lack of precipitation limits soil formation in two ways. First, little water moves through the soil profile and below the root zone, limiting horizon formation. Second, the lack of vegetation limits the amount of organic residues available to enrich the A horizon.

Major rivers, such as the Colorado in North America and the Nile in Africa, run through the desert, and throughout history have attracted settlements. To produce food, people have commonly irrigated crops on the flood plains of desert rivers. Some of these irrigation projects have been in use for hundreds, even thousands, of years. When natural flooding processes were operating, each year a new layer of sediment deposited on the flood plain, and enough water moved through the soil to

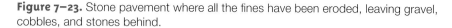

Figure 7–23. Stone pavement where all the fines have been eroded, leaving gravel, cobbles, and stones behind.

Figure 7–25. Biological, or cryptogamic, crust.

Figure 7-26. Aridsols have an A horizon that is often similar in color to the B horizon due to low inputs of organic matter. The light (white) colors on the profile are due to accumulations of calcium carbonates ($CaCO_3$).

Grasses are common in the open spaces between the shrubs. Animals are similar to those found in deserts and grasslands.

Mollisols and Alfisols are the dominant soil orders of shrublands (figure 7–2). Since soils in this biome are similar to those in grasslands, their management is similar. Although the global area occupied by shrublands is small, human activity in them has increased substantially. In California, shrublands now occur on areas that were previously part of the temperate rainforests. And ancient literature reveals that many parts of the Mediterranean coast, now covered in shrublands, were once a mixed temperate forest. Because of their comfortable climate and proximity to both mountains and coasts, shrublands are threatened by development and urbanization. Humans are also altering the natural plant succession by putting out fires, which allows shrublands to expand into neighboring grasslands.

leach the salts below the root zone. But irrigated agriculture in desert regions requires careful management to limit the buildup of salts, or salinization (chapter 6). For examples of past and current salt management issues, see chapter 8.

SHRUBLANDS

Shrublands occupy about 5% of Earth's land surface, and are found on the west coast of North America, South America, Australia, South Africa, and the Mediterranean coastal areas (figure 7–1, table 7–1). This biome[1] is characterized by evergreen shrubs that have small leathery leaves, which limit transpiration and improve the plants' fire tolerance (figure 7–27). Several aromatic herbs, such as rosemary, thyme, and oregano, come from shrublands.

Figure 7-27. Shrubland. The leather leaves of this creosote bush (inset) limit transpiration and improve fire tolerance.

---

[1] Ecologists do not agree about whether shrublands constitute a biome. Several classification systems omit it entirely, including shrublands with savanna grasslands.

AQUATIC BIOMES/WETLANDS

Water covers about 75% of Earth. This makes aquatic biomes the broadest of the biome groups. Aquatic biomes are further categorized into freshwater (rivers, wetlands, streams, riparian, lakes, and ponds) and marine biomes (coastal/estuary, continental shelf, and deep sea). The difference is the concentration of salt in the water. Freshwater biomes have less than 1% salt and close association to their surrounding terrestrial biomes. Freshwater and marine biomes intersect where fresh water flows into the sea and tides push seawater upstream into rivers. There are many differences between the two types of aquatic biomes, and the differences determine the plant and animal life in each. We focus here on freshwater and marine wetlands, in which the water is shallow enough for plants to grow above the water surface. Soils that form in the shallow water often accumulate significant organic material from the plant roots and leaves.

Wetlands are found in every climatic region and on every continent except Antarctica. An estimated 7 million to 9 million square kilometers (2.7 to 3.5 million square miles), or about 4% to 6% of the world's land surface, are wetlands (table 7–1). They form under a variety of conditions and vary in size.

Wetlands interact with both adjacent aquatic and terrestrial surroundings and therefore combine the attributes of both (figure 7–28). This interaction produces unique characteristics. Freshwater wetlands occur inland and include swamps, marshes, pocosins, bogs, and fens. Bogs and fens are also known as peat lands. Marine wetlands develop in coastal environments and include salt marshes, tidal marshes, mangrove swamps, and coral reefs.

Freshwater wetland vegetation consists of *macrophytes* and *hydrophytes* that have adapted to waterlogged, oxygen-deficient (anaerobic) soil environments. Vegetation changes from flood-tolerant tree species on wetland edges to cattails and sedges to

Figure 7–28. Wetland soils form in the presence of water or saturation in the upper 20 centimeters (12 inches) of the surface such as in this tidal marsh.

hydrophytes in the deeper portions of the wetland. Dominant species in freshwater marshes vary; however, the prevalent species are grass-like reeds and cattails, and herbaceous broadleaf plants such as smartweed and duckweed.

Bogs are peat-accumulating wetlands that have no significant water inflows or outflows and support acidophilic (acid-loving) mosses such as sphagnum. Fens, on the other hand, are peat-accumulating wetlands that receive inflows from the surrounding mineral soil and support reeds, cattails, and herbaceous broadleaf plants. Bogs can develop from mosses alone, or in combination with sedges, shrubs, and stunted trees such as pine, spruce, larch, tamarack, and crowberry.

In contrast to freshwater wetlands, marine wetlands develop in salty or brackish water in coastal regions. Coastal wetlands exist at the transition between land and ocean and are greatly influenced by ocean tides and saltwater. The presence of vegetation occurs where the land is sheltered enough from waves to permit the buildup of sediments. Coastal wetland plants, such as reed, rush, cord grass, salt grass, pond lily, and bald cypress, have developed means of tolerating a high salt content that is lethal to most other plants.

Salt marshes are found throughout the world, in middle to high latitudes along intertidal shores; in tropical and subtropical regions mangrove swamps take their place, in wet and dry tropical climates. An estimated 240,000 square kilometers (94,000 square miles) of mangrove wetlands can be found around the world. In the United States, mangrove swamps are limited to Puerto Rico and the southern parts of Florida. Their vegetation is noted for prop roots and flying buttress trees.

Like salt marshes, mangroves protect shores from wave erosion and trap sediments to build land out into the sea.

Wetlands soils are saturated or waterlogged for a considerable part of the year, or have a high water table near or above the soil surface that causes water to be retained seasonally or for long periods. The saturation of the soil renders the soil anaerobic, because the soil pores are filled with water and there is no air exchange with the atmosphere. These soils are classified as hydric, which means they possess characteristics associated with reducing conditions (figure 7–29, see chapter 5).

Histosols are by far the dominant soil order associated with wetlands (figure 7–30). They are especially prevalent in cold regions, although they can be formed in all climates where plants grow and wetland conditions exist. The main soil development process with Histosols is accumulation of organic parent material. Because of the anaerobic condition under which Histosols have developed, horizon differentiation is minimal, and when present, reflects the type of vegetation that contributed the parent material. ▷➤

Although Histosols are generally considered wetlands soils, not all wetlands have Histosols because wetlands occur on mineral soils as well (figure 7–31). Mineral wetland soils have an aquic moisture regime (table 7–3). They have colors and *redoximorphic features* (chapter 2) that indicate saturation, and reduction occurring within the upper 30 centimeters (12 inches) of the surface.

Figure 7–29. Evidence of reducing conditions includes the presence of redoximorphic features— *redox depletions* (gray) and *redox concentrations* (orange).

Figure 7–30. Histosols are wet, rich in organic material, and lack distinct horizons.

Because of the lack of oxygen, wetlands are a natural source of hydrogen sulfide, a toxic gas, and methane, a greenhouse gas. Managing wetlands presents several challenges. As population growth encroaches on wetlands in urban areas, many wetlands have been drained for construction purposes, as in the Meadowlands of New Jersey. In other places, wetlands have been drained for cultivation.

To prepare wetlands for cultivation, they first have to be drained to lower the water table below the root zone of the desired crop. The drainage has to be maintained to prevent saturation for extended periods. Since draining a wetland introduces oxygen, organic soils begin to decompose. In addition, organic soils are acidic in nature and must be limed to raise the soil pH to produce crops. Adequate fertilization also is required because the soils are low in available nitrogen, phosphorus, potassium, and micronutrients. Drained organic soils are subject to wind erosion and fire. As we learned in chapter 5, wetlands provide important ecological services, and converting them to agriculture or other uses is often controversial and is regulated by federal law.

Figure 7–31. A typical mineral hydric soil. The upper 15 centimeters (6 inches) is black due to accumulation of organic material. The gray colors are due to reduction of iron (Fe). The orange colors are where reduced iron (Fe) has been oxidized (redox concentrations).

The Great Dismal Swamp: A Pocosin, or Swamp on a Hill

The word *pocosin* comes from a Native American Algonquin word meaning "swamp on a hill," which appropriately describes the location of these wetlands.

Pocosins are evergreen shrub bogs found in the coastal plains of the southeastern United States, from Virginia to Florida. They are especially common in North Carolina. They usually occur on broad, flat upland areas between, and for the most part unconnected to, existing streams. Because of their location, they are *ombrotrophic* (rain fed) and have little to no standing water on their surfaces. Some may be fed by seepage or springs. If the organic soil is removed to a depth below the shallow water table and down to the mineral material below, a Pocosin lake will form. They vary in size from a hectare to thousands of hectares (about 2.5 acres to thousands of acres), and those that have remained undisturbed provide a refuge for wildlife.

Pocosins typically have vegetation consisting of pond pine, loblolly pine, longleaf pine, holly, bay, Venus's-flytrap, and sweet pitcher plant. Because pocosins periodically dry out in spring or summer, fires occur every few decades. The fires can be intense and burn deep into the moss, creating a pond or lake.

Pocosins have nutrient-deficient, acid soils (sandy or clay-based) that are saturated for much of the year or shallowly flooded primarily during cool seasons. They are not alluvial, but have upland landscape features. Since these occur on broad flat landscapes external and likely internal drainage was slow and the water table was high to begin with. Under these saturated and often anaerobic conditions annual leaf litter decomposes slowly to muck (plant fibers are no longer visible due to decomposition). The muck has a low hydraulic conductivity, causing the water table to rise higher. The result is a positive feedback that causes the surface to rise as organic matter builds up due to slow decomposition. The slow conductivity of the organic soil, coupled with its high reactivity result in the water leaving the pocosin being purified and slowly released to estuaries. This is important as it helps maintain the proper nutrient, acidity, and salinity in the estuary.

SUMMARY

SOILS ARE AN INTEGRAL PART OF EVERY TERRESTRIAL BIOME. You may have realized the answer to the question, "Which came first, the soil or the biome?" The answer is, they developed together. Changes in one, perhaps over hundreds or thousands of years, affect the other. Just as the living portion reacts to the abiotic factors of the biome, so does the soil respond to the biotic factors. Thus, the soil and biome are intricately linked to one another.

CREDITS

7-1, P. Reich, USDA-NRCS; 7-2, USDA-NRCS; 7-3, 7-4, P. Reich, USDA-NRCS; 7-5, D. Weindorf; 7-6, USDA-NRCS; 7-7, D. Lindbo; 7-8, V. Da Silva Fraga and I.H. Salcedo; 7-9, D. Lindbo; 7-10, 7-11, 7-12, D. Lindbo; 7-13, J. Vanuga, USDA; 7-14, USDA-NRCS; 7-15, USDA-NRCS; 7-16, T. Robertson; 7-17, USDA-NRCS; 7-18, D. Weindorf; 7-19, A. Jones, Ph.D./Flickr; 7-21, D. Lindbo; 7-22a, Morguefile; 7-22b, NSF; 7-23, C. Robinson; 7-24, 7-25, 7-26, D. Lindbo; 7-27, C. Robinson; 7-28, morguefile; 7-29, 7-30, 7-31, D. Lindbo; fire sidebar photos, C. Robinson. Chapter opener image, iStock.

Copyright © 2012. Soil Science Society of America, 5585 Guilford Rd., Madison, WI 53711-5801, USA. *Know Soil, Know Life.* David Lindbo, Deb A. Kozlowski, and Clay Robinson, Editors
doi:10.2136/2012.knowsoil.c7

GLOSSARY

ABIOTIC FACTOR A nonliving condition or parameter (soil, sunlight, climate, landscape, geology, aspect, latitude, water) that influences or affects an ecosystem and the organisms in it.

AGROFORESTRY Any type of multiple cropping land-use that entails complementary relations between tree and agricultural crops and produces some combination of food, fruit, fodder, fuel, wood, mulches, or other products.

ALLEY CROPPING Growing vegetable, grain and/or forage crops between rows of trees.

ASPECT The compass direction that the slops faces in relation to the sun.

BIOME A large geographic region with similar environment and distinctive plant and animal community.

DEFORESTATION The permanent removal of trees until less than 10% of the forested land remains.

DESERTIFICATION Land degradation where the land area is becoming increasing dry and inhospitable to plants and animals.

DIURNAL Any pattern that recurs daily.

ECOSYSTEM A system formed by the interaction of a community of organisms with their environment.

EPIPHYTE A plant that grows on another plant and depends on it for support but not food (non-parasitic). Epiphytes get moisture and nutrients from the air or from small pools of water that collect on the host plant. Spanish moss and many orchids are epiphytes. This is an example of commensalism.

FALLOW Tilled but not planted to a crop during the growing season.

HYDROPHOBIC SOILS (HYDROPHOBICITY) The tendency for a soil particle or soil mass to resist hydration, usually quantified using the water drop penetration time test.

HYDROPHYTE Plants that are adapted to grow in water or wetlands.

IMMOBILIZE (IMMOBILIZATION) Conversion of plant available nutrients to an organic form which is unavailable to plants.

MACROPHYTE Wetland vegetation that may grow above the water surface (emergent), remain below the water surface (submergent), or float on top of the water.

OMBROTROPHIC A bog, soil or vegetation that receives all moisture from atmospheric sources such as precipitation rather than groundwater, springs, or rivers.

PERMAFROST Permanently frozen subsoil, occurring throughout the polar regions and locally in perennially frigid areas.

REDOX CONCENTRATIONS Zone where iron oxides have accumulated due to oxidation of reduced iron—high chroma mottles.

REDOX DEPLETIONS Zone where iron oxides coatings have been removed due to reduction of iron—low chroma mottles or matrix.

REDOXIMORPHIC FEATURES Features or morphology related to the reduction, oxidation and translocation of carbon, iron, manganese, and sulfur.

SAPROPHYTE Any organism living upon dead or decaying organic matter.

SHIFTING CULTIVATION Agricultural systems similar to long-term rotations, in which areas are cultivated for a few years until nutrients are depleted, then abandoned to be reclaimed by native vegetation. Cultivation moves to an adjacent area. Native vegetation restores some fertility so the area can be cultivated again ten or more years later.

SOIL MOISTURE REGIMES Refers to the moisture content in the upper portion of the soil.

- **Aquic:** soil is saturated and reduced near the soil surface for a portion of the year
- **Aridic:** soil has little to no available water for plant growth for a substantial portion of the year
- **Udic:** soil is neither dry nor wet for long periods during the year
- **Ustic:** soil has a dry period but not as dry a aridic
- **Xeric:** soil has moist cool periods and warm dry periods.

SOIL TEMPERATURE REGIMES Refers to the average annual soil temperature at 50 centimeters (20 inches) below the soil surface.

- **Gelic:** average annual soil temperature is near or below freezing, has permafrost
- **Cryic:** average annual soil temperature soil is near freezing, lacks permafrost
- **Frigid:** average annual soil temperature soil is near freezing but has warmer summer temperatures than cryic
- **Mesic:** average annual soil temperature is well above freezing, sometimes referred as the corn belt
- **Thermic:** average annual soil temperature is higher than mesic, sometimes refered to as the cotton belt
- **Hyperthermic:** average annual soil temperature is high than thermic, sometimes referred to as the citrus belt

TAIGA A biome characterized by coniferous forests consisting mostly of pines, spruces, and larches.

TERMITARIA Termite mounds.

CHAPTER 8

SOILS AND SOCIETY

CHAPTER AUTHORS

MELANIE SZULCZEWSKI
MANDY LIESCH
DAVID LINDBO
JOHN HAVLIN

"Civilization itself rests upon the soil," wrote Thomas Jefferson, yet now, in many ways, the fate of our soils rests with civilization. For thousands of years, societies have been using natural resources and changing their environment to provide themselves with food, water, housing, and many other goods. As we've seen in this book, soils are essential to agriculture, feeding the world's population, and sustaining all life on Earth. Soils also provide countless other products besides food, directly (as with clay or potting soil) and indirectly (such as timber and rubber). As the human population continues to increase and societies continue to expand use of natural resources, properly caring for the soil becomes more and more important.

This book has focused on our reliance on soil, but it is important to note that humans have also recognized the beauty and importance inherent to soils and have relied on them for spiritual, artistic, and literary inspiration. This chapter looks at the role that soils play in human culture. You'll learn about societies past and present and that whether they thrive, flounder, or fail depends on whether they care for the soil.

SOILS AND HUMAN CULTURE
SOIL AND ART

As described earlier, soil is made up of sand, silt, and clay particles. All parts of the soil serve human needs, but of these three types of particles, clay has probably played the most important role for societies. Most of the materials used in the making of traditional artwork are found in the earth's crust. These include soil for color pigments used in paints, dyes, and inks; rocks for sculptures; sand and minerals to create glass; ores for metalwork; and clay for ceramics. Soils are also critically important, as we've seen throughout this book, in growing plant materials for textiles and wooden objects.

In the Stone Age, natural pigments like red and yellow ochres and magnesium oxides were used for pottery, houses, and body paint for rituals (figure 8–1). Burial *cairns* covered in red ochre have been found in Europe, Japan, and North America. For centuries, the Bamana people of Mali have been making "mud cloth" for clothing as an important symbol of their culture and heritage (figure 8–2). To make mud cloth, white cotton is washed and shrunk. The white cloth is coated with leaves, generating a yellow color. Mud is collected from ponds and other areas and kept in a pot for up to a year. The cloth is then painted with elaborate designs using many different colors of stored mud. A natural plant chemical is then applied to the cloth to make it colorfast, turning the yellow areas white and leaving a black pattern against a white background. Traditionally, the patterns held significant meanings and purposes, such as protection for the wearer.

The Dutch painter Hieronymus Bosch (1450–1516) used earth and soils as the main subjects of his paintings, to evoke deep religious symbolism and meanings. His most famous and controversial piece is the triptych *Garden of Earthly Delights* (figure 8–3), which portrayed people living off the bounty of the earth.

Figure 8–1. Neolithic Europeans used soil pigments, such as red ochre, to decorate the inside of caves, as with this painting of a horse in France.

Figure 8–2. The Bamana people of Mali use many different soils as cloth dyes to create African mud cloth.

Figure 8–3. *The Garden of Earthly Delights* by Hieronymus Bosch is an early example of people living off the bounties of the soil, as they did in the Garden of Eden. Like most paintings of the time, the third panel on the right hand side shows a hellscape of the consequences of a life filled with nothing but indulgence.

Figure 8–4. Soils have multiple layers and colors, making them beautiful works of art by themselves.

Soil is well represented in art through agricultural scenes starting in the 1500s, but it was not the exclusive subject of a body of work until the 1950s. This environmental art movement is known as the Land Art Movement. In these paintings, soils are the focus. Soil profiles can show stark contrasts in colors, layers, and textures, making for beautiful art (figure 8–4)

An artist from California, Gary Simpson (born 1947), took on an ambitious soils project entitled *Common Ground 191*. When this project is complete, he will have collected soils from 192 countries to create panels that represent the different parts of the world (figure 8–5). The major theme is that all countries may be different, but they are still united, piece-by-piece, as a whole earth. Another contemporary artist, herman de vries (born 1931), created an exhibit of more than 7,000 soil samples from all over the world, which is displayed at the Gassendi Museum in southern France. The different colors of soils are the primary art form (figure 8–6).

Figure 8–5. *Common Ground 191*, by Gary Simpson, incorporates soils from 192 countries into one cohesive piece.

Figure 8–6. *from earth: stiftsland*, by herman de vries, represents many different soil colors, with the soil itself as artwork.

Figure 8–7. The Code of Hammurabi, which was written on clay tablets, held all of the important laws in ancient Babylon. The tablets relay 282 rules about how people should behave in Ancient Babylon. These rules had consequences for violations, including the well-known "an eye for an eye, a life for a life" punishment.

SOIL AND LITERATURE

As with art, soil can serve as the subject of a written work, act as a powerful literary image, or provide the very canvas for writing itself. A lot of the permanent writings from early societies are preserved on clay tablets (soil), rather than stone. The tablets were written on while wet with sharp objects and then fired into a permanent record of political laws, trading documents, and ideas. The Code of Hammurabi from 1750 BC (figure 8–7) is one of the oldest remaining examples of written law on clay tablets.

Writers have used soil as a metaphor for birth, death, the life cycle, fertility, and rejuvenation for centuries. Many scientists, including Isaac Newton and Benjamin Franklin, have written about soils in their poetry and personal writings. One famous stanza is from Franklin's *Poor Richard's Almanack* (1739):

> A MAN OF KNOWLEDGE LIKE
> A RICH SOIL, FEEDS
>
> IF NOT A WORLD OF CORN,
> A WORLD OF WEEDS.

In addition to studying the evolution of species, Charles Darwin had a lifelong interest in the evolution of soil. He was the first to recognize the beneficial effects of earthworms on soil formation and fertility, noting in his popular 1883 book, *The Formation of Vegetable Mould Through the Action of Worms with Observations of Their Habits*, that

> I WAS THUS LED TO CONCLUDE, THAT
> ALL THE VEGETABLE MOULD [SOIL]
> OVER THE WHOLE COUNTRY HAS PASSED
> MANY TIMES THROUGH, AND WILL
> AGAIN PASS MANY TIMES THROUGH,
> THE INTESTINAL CANAL OF WORMS.

Many 19th and 20th century poets used soil to enhance the imagery of their writing. Walt Whitman wrote in "Song of Myself" (1855):

> MY TONGUE, EVERY ATOM OF MY BLOOD,
> FORM'D FROM THIS SOIL, THIS AIR.

Robert Frost, one of America's most famous 20th century poets, used imagery of soil *erosion* to begin his poem, "In Time of Cloudburst" (1936):

> LET THE DOWNPOUR ROIL AND TOIL!
> THE WORST IT CAN DO TO ME
> IS CARRY SOME GARDEN SOIL
> A LITTLE NEARER THE SEA.

In 20th century Russia, many influential poets, including Gleb Gorbovsky, were soil scientists and other earth scientists. These poets are referred to as *pochveniks* or "soil-heads."

Soil plays a starring role in some novels, including Bram Stoker's *Dracula* (1897). In order to rest properly, Dracula needed soil from his Transylvanian home, so he had 50 boxes of soil shipped to himself in England (figure 8–8). The famed vampire hunter,

Figure 8–8. Vampires, such as Dracula, often have ties to the soil and need to sleep on native soil every day, or they cannot feed.

Van Helsing, discovered this need for native soil and placed communion wafers on 49 of them, repelling Count Dracula by their holy nature. This book marked the beginning of modern vampire stories, and since then, vampires have had a powerful connection to soil in popular culture.

John Steinbeck gave soil a prominent role in his books. *The Grapes of Wrath* (1940) examined American life during the Dust Bowl. When small farmers were unable to pay mortgages or buy equipment like tractors to till fields, many families were forced to leave their homes and travel westward to California, looking for work (figure 8–9). Another of his novels, *East of Eden* (1952), relayed the difficulty of people trying to eke out an existence on infertile land.

Real life soil scientists can also inspire science fiction writers, like Frank Herbert, who wrote the novel *Dune* (1965). Herbert was researching a U.S. Department of Agriculture (USDA) dune stabilization and restoration project in Florence, Oregon for a magazine article and became so inspired by this research he incorporated the ideas into his fictional universe. The character, Pardot Kynes, who is an ecologist, geologist, and meteorologist, is reportedly based on the work of a USDA soil scientist.

SOIL AND WARFARE

Art and literature have reflected the importance of soils throughout the centuries, often resulting in the best manifestations of our culture. The impact of soils is evident in the worst manifestation, the human institution of war. Most of the battles fought throughout the course of history have been fought on land—in other words, on soil. Soil is not often noticed in battle strategy, but its properties have been critical to the success of certain battles and maneuvers.

The battle of Agincourt (1415) may best be known for the stunning defeat of a vastly superior French force by the English through the use of the longbow. Interestingly, the terrain where the battle took place was equally as important. The fields along the English front line had recently been plowed and were wet from recent rains. As the French advanced the fields turned to mud. Soldiers in their heavy plate armor sank to their knees in the mud. This slowed the advance, making them vulnerable to the barrage of English arrows, and prevented the French from moving out of the way of both the English and their own troops approaching from behind. In the end the soil may have been just as decisive as the rain of arrows.

Mud has continued to plague military operations throughout history. Mud was the bane to all involved in the trench warfare of World War I. During one battle, fighting around a reclaimed marshland in Belgium turned ugly. Heavy bombing, soils that were already soaked, and torrential

Figure 8–9. The Dust Bowl in the 1930s caused many farmers to lose their land. These "Okies" packed up all of their belongings and headed west to find new opportunity.

Figure 8–10. In the Third Battle of Ypres, fought in Belgium during World War I, artillery shell explosions churned up the heavy clays, which engulfed the cannons with every fired shot.

Figure 8–11. The North Vietnamese developed an extensive network of tiny, underground tunnels in their soils. These entries were well camouflaged, and so small, that most American troops could not enter. The tunnels are now open to tourist groups.

rains turned the battlefield into a large, deep mud pit (figure 8–10). These large pits of mud also made it nearly impossible to get wounded men to medical facilities. Wounded men drowned in the liquid mud. Mud also severely delayed the German movement into the Soviet Union during World War II.

In the Vietnam Conflict, the North Vietnamese army made use of a centuries-old network of paths through the mountains called the Ho Chi Minh Trail. The amount of troops, food, and weapons transported along the trail had to be greatly reduced, however, during the rainy season due to boggy, muddy conditions. The U.S. military developed a strategy to take advantage of this predicament by seeding clouds to encourage rain and extend the monsoon season. In response, the North Vietnamese developed a network of underground tunnels (figure 8–11) that they lived in and used for transportation of goods. These tunnels were like caves and still exist today because of the types of soils present in the subsurface.

Mud is not the only problem during war; dust from dry soil can affect maneuvers on land and in the air. The Carthaginian general Hannibal observed the weather patterns around Rome and "fought with wind, dust, and sun in his favor," defeating the great empire with the weapon of dry soil, according to second century historian Lucius Annaeus Florus. Dust can hide entire armies and targets in large clouds, as well as alert sentries of an approaching enemy. Dust also damages machinery. A tank engine's life span is cut in half in the desert.

SOILS AND HUMAN HEALTH

The most direct connection between soil and human health is in the soil's ability to provide the nutrients we need to thrive. Soil also plays other roles in medicine and disease.

Soil contains between 50 billion to 1 trillion microorganisms in every tablespoon, and most belong to as yet unidentified species (see chapter 3). Microorganisms produce many chemicals in their associations with each other. Because of the large variety of chemical-producing bacteria, fungi, and other organisms, soil has the potential to function as a major pharmaceutical supplier. Unfortunately, soil also has the potential to harbor disease-causing *pathogens*.

THE DANGERS

Children are often told that soil is dirty, and that they should always wash their hands after playing in it. There is merit to this, as worldwide several million people per year are infected with and made seriously ill by soil-borne pathogens. Soils may harbor parasites, such as toxoplasma, hookworms, and roundworms, and several varieties of fungal and bacterial pathogens that can cause anthrax, botulism, tetanus, or other diseases. These pathogens, like flu viruses and many other pathogens, often disproportionally affect those in weak health or with vulnerable immune systems. Ironically, one theory for the rise in the number of children with allergies proposes that many are raised in environments that are too sterile—their lack of exposure to soil and its biological components prevents their immune systems from developing properly.

THE BENEFITS

Because soil has a large number of biochemically versatile bacteria and fungi, it also provides antidotes for many maladies (chapter 3). The *antibiotic* streptomycin was discovered by soil microbiologists Albert Schatz and Selman Waksman and is one of the many medicines produced by soil microbes. Waksman's epitaph reads, "The earth will open and bring forth salvation," suggesting his appreciation for all that soil can give us.

Some soil bacteria also produce compounds that show promise in treating tumors, and a soil fungus is the source of cyclosporine, a drug widely used to prevent transplant patients from rejecting their new organs. Soils provide base ingredients for antidiarrheal drugs, such as kaolin and pectin for antacids. There are also mud therapies (called *thermalism*) that people with dermatitis, diabetes, and arthritis have used for centuries. The immense diversity of soil microorganisms and other soil components are still being explored for treating infections, cancer, and even neurological issues.

GEOPHAGY: EARTH EATING

Despite what we know about pathogens in the soil, people sometimes deliberately eat soil or clay, a condition known as *geophagy*. Some children may go through a soil-eating phase, but in adults, the eating of soil generally is considered abnormal. In some cases, people are so hungry they eat soil for nutrients and to fill their stomach. Sometimes geophagy results from the physiological craving for calcium and iron. In other cases, the geophagy is cultural.

Geophagy has even produced a successful business: one Nigerian village produces more than 500 tons of soil per year for consumption across other regions in West Africa. Geophagy also occurs in urban South Africa, parts of the southeastern United States, and the Middle East. In South Africa eating clay is believed to soften the skin, making one more attractive. In some areas, clay is used as a preventative traditional "medicine" before eating potentially toxic food since it is believed to "line the stomach." Some believe that soil can reduce morning sickness in pregnant women and provide necessary micronutrients that the local diet may be missing. While there may be merit in these beliefs, the risks outweigh the unproven benefits.

It is important to realize that geophagy can be dangerous. People ingesting soils can acquire parasitic infections and heavy metal poisoning (from lead, cadmium, or arsenic). Soils with very high potassium content can cause cardiac arrhythmia or arrest. Ingesting soils with high cation exchange capacities (chapter 4) can result in iron deficiencies. In general, it is safe to garden and play in the soil, but everyone should acquire the habit of washing their hands afterwards and definitely before you eat.

CHALLENGES TO THE SOIL: THEN, NOW, AND LOOKING AHEAD

No civilization can last forever. There are always several reasons contributing to the collapse of a thriving society, but historically most crumbled because of environmental, political, or social problems that led to the destruction of their soil. Growing populations and their demand for food put pressure on the soil. People remove trees or vegetation from an area with devastating effects on the soil, including erosion and fertility loss. The soil may become exhausted over time and run out of natural fertility. Irrigation systems fail, canals silt up, or salt accumulates. All of these environmental issues affect the soil in numerous direct and indirect ways, generating social pressures and problems. Numerous past and present societies have had to cope with challenges to their soil. Fortunately, our modern civilization has learned many lessons from the past, and many people and organizations are implementing solutions to maintain and restore the health of our soils.

THE CHALLENGE OF DEFORESTATION

Deforestation can have destructive effects on soils, as we learned in chapter 6. Recall that *deforestation*, as defined by the United Nations, is the permanent removal of trees until there is less than 10% of the forested land remaining. Since the beginning of the Industrial Revolution, humans have removed more than half of the original forest cover on Earth. Each year, an area the size of Greece is deforested (see chapter 6). Humans cut down trees to build houses and ships, to develop towns, and to clear land for agriculture. Trees also provide an important fuel source and continue to serve this purpose for at least 2 billion people. Wood is a form of renewable, carbon-neutral energy when managed correctly, but managing and harvesting timber improperly can lead to deforestation and other environmental issues, including the degradation of soil.

Many people in developing countries today practice a type of farming called slash and burn agriculture. After a section of forest is cut and burned, an open area with soil full of nutrients is left for growing crops, since the ashes from the burned vegetation contain all the nutrients that were in the plants. This technique temporarily increases the fertility of a soil, but the effect does not last, especially in tropical areas. Tropical soils tend to be nutrient poor, as the typical lush vegetation contains most of the nutrients available in this ecosystem. Since each successive crop removes more nutrients from the soil and none are replaced, farming is only successful for a few years.

This method of slash and burn agriculture can be sustainable with small communities, if the forest is allowed to regrow while the farmers move on to a different plot of land. Today's increased demand for food and *cash crops* requires land to be used repeatedly, however, without time for recovery, and with each harvest, the soil becomes less and less fertile.

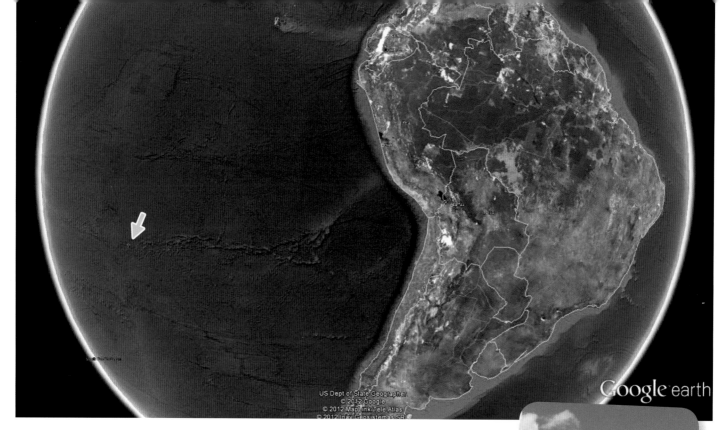

Figure 8–12. Easter Island, perhaps best known for its large statues called moa, is a remote location off the coast of Chile. When they first settled on the island, the colonists had plentiful resources; however, being so isolated, and without trade, they eventually depleted their resources, resulting in a failed civilization.

HISTORICAL CASE: EASTER ISLAND

The setting sun silhouettes some 600 massive stone statues, averaging more than 20 feet high, on Easter Island, a small remote island in the middle of the Pacific Ocean (figure 8–12). When Polynesian explorers arrived there, dated by some as about 1,000 years before European discovery, they found an island with adequate rainfall and decent soil.

The only crop that grew well on the island was the sweet potato, which required very little time and effort to grow. Some anthropologists believe that perhaps because of this, the islanders had a lot of time to dedicate to building the impressive stone statues. Pollen and other scientific evidence indicate that an ample number of trees were on Easter Island in the early days of the fifth century when the population was relatively small. The islanders cleared trees for farms and also for heating, cooking, building homes, and creating canoes for fishing. They also used the trees to help roll the statues from one location to another.

By 1550, the population had grown to at least 15,000 people. With so many islanders, there was more competition for resources, resulting in the development of a number of clan groups, which competed with each other to acquire resources and build the grandest statues. This led them to cut down more trees, which greatly increased soil erosion, decreasing soil fertility. By about 1650, wood on Easter Island ran out, and the food began to run out too because the soils were so badly eroded. The islanders did not realize that they needed to replenish nutrients in the soil, and even if they had, they lacked nutrient sources, such as manure from domesticated animals, on the island. Centuries of deforestation intensified soil erosion, removing the topsoil in a short period of time.

Recent studies have turned up evidence indicating that severe sheet erosion occurred between about 1250 and 1500, corresponding with the period of intensive agriculture and land clearing. Crop yields declined suddenly, and starving people engaged in warfare. When Dutch explorers discovered the treeless island on Easter morning in 1722, they found about 2,000 poverty-stricken inhabitants, living in caves and reed huts. A great society had fallen because of the effects of deforestation, soil loss and loss of productivity, and uncontrolled population growth.

MODERN CASE: HAITI

Haiti is considered to be the poorest country in the Western Hemisphere and one of the poorest countries in the world. This unfortunate ranking can be directly linked to the degradation of Haiti's environmental resources. Before colonization, the country was covered with trees; now, less than 1% of the original tree cover remains. The country is also relatively mountainous, so most agricultural land is prone to erosion. Like the Easter Islanders before them, the Haitian population is rapidly growing, from 3 million people in 1940 to 10 million in 2010, requiring more and more food and resources. However, the soils are so degraded and infertile, that it is difficult to generate any more food.

Most Haitians are subsistence farmers, forced to clear more mountainsides to grow more food. The farmers also need the trees as a source of fuel and cash. People living in Port-au-Prince, the capital of Haiti, are in dire need of charcoal as an energy source. Many trees must be cut down and slowly burned in kilns to make charcoal, which is easily transported to urban areas. For rural people, providing charcoal to the cities accounts for nearly 20% of their income. This leads to rampant deforestation, especially on the mountain slopes (figure 8–13). The result is dramatic soil erosion, an estimated 6,000 hectares (14,862 acres) of soil lost per year. When it rains, or there is a hurricane, these soils have nothing to hold them in place, resulting in landslides, mudslides, and even more severe flooding, which further erodes the soil.

Figure 8–13. Haiti shares the island of San Salvador with the Dominican Republic, which is covered in trees (right), in comparison with the Haitian side (left). Without deep-rooted trees to hold the Haitian soil, the soil erodes with the rain.

LOOKING AHEAD: TACKLING RAINFOREST DESTRUCTION IN BRAZIL

Brazil today has the world's highest annual rate of deforestation. As a result, the populous country faces a loss of 20 tons of soil per hectare per year, with 2 billion tons muddying the Amazon and other rivers annually. Brazil saw the productivity and *biodiversity* of its ecosystems plummet, so the Brazilian government stepped in to reverse the forest destruction. National laws now protect more than 110 million hectares of rainforest, representing the largest safeguarded area in the world. Such policies have resulted in a 40% decrease of annual deforestation since 2004. While this indicates an admirable effort, the Amazon rainforest is still disappearing at an alarming rate. As the thin layer of productive soil washes away, the sediment load in the rivers increases, damaging aquatic life downstream.

Since large-scale forest removal is now restricted, most of the current land clearing is due to localized slash and burning for cattle ranching and subsistence farming. Monitoring has increased, and both government agencies and nongovernmental organizations are working together, using technology to monitor the forests. Aerial photography and satellite imaging keep track of holes in the forest and sediment loads in the waters (see figure 8–14).

Brazil's National Institute of Space Research released satellite images in 2011 indicating that deforestation had actually increased since 2010. The alarmed government used this information to focus enforcement activities in areas suffering from illegal logging and land clearing. This enforcement showed results, and the Brazilian government estimates that the deforestation rate for the last half of 2011 was the lowest ever recorded. This is good news for the biodiversity of Brazil's rainforests, with benefits extending to the nation's soil and freshwater health as well.

Figure 8–14. In Brazil's agricultural areas, erosion from deforestation (Verde River) results in muddier water than in those areas surrounded by tree cover with native swamps (Parana River).

THE CHALLENGE OF DESERTIFICATION

Desertification is the rapid loss of topsoil and loss of plant life on productive land in arid and semi-arid regions (chapter 6). One third of Earth's land area in more than 100 countries is at risk of desertification. Affected regions are found throughout the world, including in the western United States, Australia, sub-Saharan Africa, parts of the Middle East, China, and Central and South America (figure 8–15). Even though desertification can happen naturally, most cases are caused by human activities: over-cultivation, vegetation removal, poor irrigation practices, and overgrazing. In arid and semi-arid lands, the soil is already fragile, and when agricultural practices without proper management damage or degrade the soil, it quickly becomes barren.

HISTORICAL CASE: THE DUST BOWL

When the High Plains of Texas, Oklahoma, Kansas, and Colorado were settled in the decades after the Homestead Act of 1862 was passed, the ranches and farmsteads enjoyed abundant rainfall. However, a 10-year drought, from the early 1920s to 1930s, followed the overgrazing and excessive tillage that had been encouraged to meet food needs during World War I. Crops failed in the drought, and when winds came, the fertile soils went with them. People wore dust masks and hung wet sheets over windows to keep out the dust. The towering dust storms generated by these conditions were called *black blizzards* (figure 8–16).

Many farmers lost all their money, and farms failed. By 1934, the drought covered 75% of the country. It was estimated that 100 million acres had lost most of

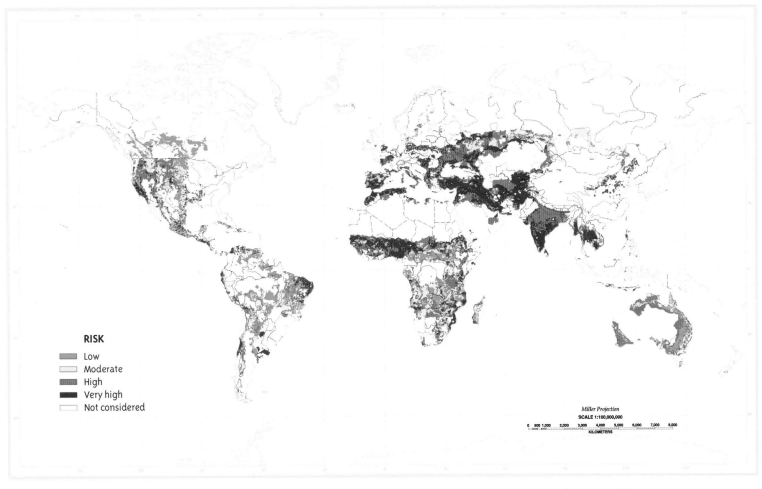

Figure 8–15. Areas most vulnerable to human-induced desertification are found throughout the world, affecting 1 billion people.

their topsoil. Soil rained down as far east as Washington, DC, so the federal government declared erosion a menace, created the Soil Conservation Service (now the Natural Resource Conservation Service), and implemented federally mandated soil and range conservation areas. These programs are still active today in protecting soils from erosion and other degradation, although some Dust Bowl lands are still at high risk for desertification.

MODERN CASE: THE DISAPPEARING ARAL SEA

The Aral Sea, located between Kazakhstan and Uzbekistan, was once one of the largest lakes in the world, but more than 80% of it has disappeared in the past 50 years. Starting in the 1960s, the Soviet Union diverted water from the Aral into irrigation canals to grow cash crops like cotton, transforming the semi-arid region into a productive agricultural area—temporarily. The cotton farming disrupted a delicate soil ecosystem and decimated the lake and surrounding region. While the devastating effects on the lake ecology and fishing industry are well known, the problems facing the soil are just as severe.

As the area of the lake shrank over the years (figure 8–17), more and more of the lakebed became visible. The strong, dry winds in the region picked up the newly exposed sediments, blowing dust and salts over the surrounding land. The dust contained pesticides, fertilizers, and other chemicals that then contaminated the soil and vegetation. Without the mitigating effect of the large sea, the regional climate became more arid, increasing each of these

Figure 8–16. This black blizzard during the Great Depression is a powerful example of wind erosion and desertification.

problems. The United Nations claimed that by 2001 more than 45% of Uzbekistan's soil had been degraded, with crop yields only one-third of the historical production. As crop yields declined and fewer native plants grew in the new conditions, more water and chemicals were needed to grow crops, further draining and destroying the Aral Sea. The lake shrank even more, more toxic dust blew around, more soil was contaminated and degraded, and the cycle worsened until the once productive area became a dry barren wasteland.

The local communities are now dealing with the result of the desertification of the Aral region. Unfortunately, the southern half of the sea, now called the Big Aral, has been separated by an exposed land bar and little can be done to restore it. It is expected to completely disappear by 2020. The World Bank and some nongovernmental organizations are putting effort and millions of dollars into the recovery of the Small Aral in the north, with some positive results. In the south, people must adapt to the more desert-like conditions. Community projects are planting large tracts with the saxaul bush (figure 8–18), a salt- and drought-resistant plant, to slow the extent of desertification and restore some health to the soils.

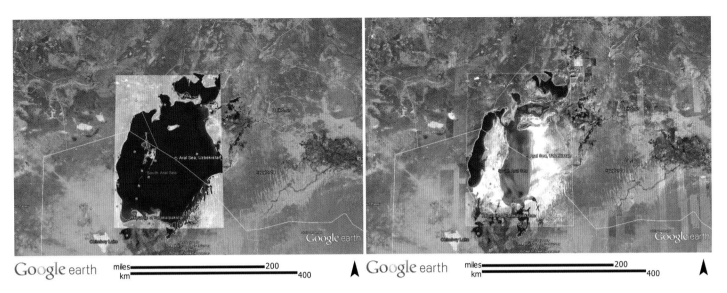

Figure 8–17. As a result of mismanaged irrigation and intensive cash crop agriculture, the Aral Sea has shrunk from one of the largest lakes in the world (left, 1973) to a dried lakebed (right, 2011) containing high quantities of salt. Without water to wet the air, the soil blows away, expanding this already dry region into a desert.

Figure 8–18. The drought- and salt-tolerant saxaul bush is being planted to combat desertification in the Aral Sea region.

LOOKING AHEAD: PLANTING A GREEN BARRIER IN AFRICA

In the dry regions of the Sahel, the region bordering the southern parts of the Sahara desert, desert-like conditions exist where productive land once supported many people. Land degradation here is mostly caused by wind erosion, as decades of extensive tree cutting, overgrazing, and droughts have reduced the amount of vegetation and roots holding the soil together. Winds blow and transport the fragile soils elsewhere. To mitigate this, the African Union developed the Great Green Wall Project. This project aims to replant and reforest this region and reverse the relentless pace of desertification in Africa. The plan is to work with local communities to improve the soil and regional climate by planting a wall of trees at least 15 kilometers (9.3 miles) wide across 11 countries (figure 8–19).

The West African country of Senegal is a leader in the project, planting an average of 2 million trees a year since 2008. New trees increase biodiversity, which enhances soil biology and rejuvenates the soil, while greatly decreasing soil erosion. These trees will also help recharge the water table by increasing soil water retention, as well as block the blowing sand grains from landing on farmland. The tree planters in this project are local volunteers, driven to improve their country and to fight poverty (figure 8–20).

THE CHALLENGE OF SALINIZATION

Arid and semi-arid lands are subject to another devastating challenge to their soil's productivity: *salinization* (see chapter 6). The natural growth potential of plants in these regions is limited by the lack of water, even though the soils can be very fertile. To use these soils, farmers and governments invest in irrigation technologies. Irrigation can increase crop yields by up to 10 times, but without proper management irrigation can turn soil into a salty, degraded mess (figure 8–21). About 20% of the total amount of irrigated land on the planet has already been damaged by salinization.

There is also a type of salinity known as dryland salinity, which does not occur on irrigated lands. Salts come from deep underground as the water table rises. This makes it difficult to predict.

High salt concentrations are not good for growing plants. Salts do not allow plants to grow or absorb moisture; they also remove moisture already in the plant. Salt also destroys the soil structure, so water cannot easily flow through it, which can cause plants to die from waterlogging. Each year more than 1.5 million hectares (3.7 million acres) of once-arable land becomes unproductive from salinization, leading to an $11 billion loss in the world food market. Many areas of the world suffer from rising salinity today.

Figure 8–20. Farmers in Senegal often come many miles to help plant trees, while local farmers protect the trees from their livestock.

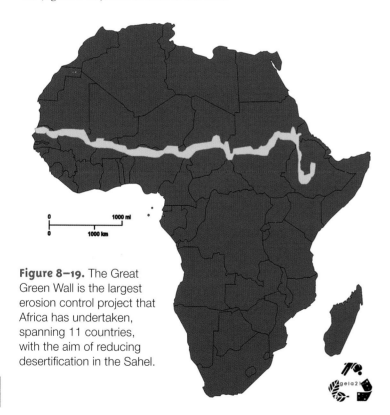

Figure 8–19. The Great Green Wall is the largest erosion control project that Africa has undertaken, spanning 11 countries, with the aim of reducing desertification in the Sahel.

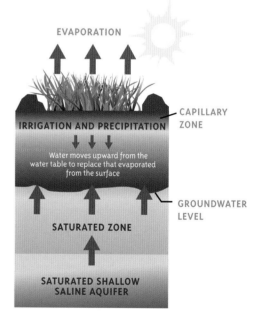

Figure 8–21. In hot and dry regions, fields need to be irrigated to best grow plants, resulting in high salinity. Adding more water to the soil increases evaporation. Water from the saturated soil, or the saline aquifers, then moves upward to take the place of the water being evaporated, bringing with it the salts from natural rocks. When this water evaporates from the surface, the salts are left behind.

HISTORICAL CASE: MESOPOTAMIA

Conditions in early Mesopotamia (now southern Iraq) were a lot like they are there today: hot and dry. The Tigris and Euphrates Rivers provided water from the mountains for this early civilization. The soil in this region is very fertile, and the ancient Mesopotamians built irrigation projects to bring water into the dry desert region. Unlike the Egyptians, who could rely on the seasonal flooding of the Nile and irrigate accordingly, the Mesopotamians dealt with rivers that were more unpredictable. Despite this, the people were able to adapt and a massive civilization developed.

This region used to be the breadbasket of an empire, producing large quantities of wheat thanks to irrigation. The region boasted a strong labor force, both agricultural and urban. Farming thrived in this region for 3,000 to 4,000 years before things started to change. The soils eventually accumulated dangerous amounts of salts because of low permeability and low precipitation. Over-irrigation caused the water table to rise, bringing up more salts, turning the fields white. Crop yields declined, and the cities did not have enough food. Historians believe that the rising soil salinity caused farmers to switch from wheat to more salt-tolerant barley between 2100 and 1900 BC.

By 1800 BC, even barley could not grow because of the ever-increasing accumulation of salts. The priests, administrators, merchants, and soldiers could not have the food they needed to practice their crafts. The cities fell into unrest when people could not get enough to eat, and many moved elsewhere. Between 2100 and 1700 BC, three-fifths of the Sumerian population of Mesopotamia moved north. Many of the great Sumerian cities never recovered, and only their ruins exist today (figure 8–22).

Figure 8–22. The City of Ur has very little remaining of its former glory except this ancient ziggurat (shrine).

MODERN CASE: AUSTRALIA

The continent of Australia is home to unique vegetation and wildlife that has adapted to small amounts of rainfall and a high salt content in the weathered rocks. Once Europeans settled in Australia, much of this native vegetation was cleared for crops, which need regular supplies of water. More water in the soil from irrigation meant that the water table rose, and with it came the salts that weathered from dissolved rocks (figure 8–23), creating what are known as saline seeps.

In some parts of Australia, the groundwater does not move quickly and annual precipitation rates are low, so trying to fix these seep regions may take several hundred years. The problem affects not only Australia's marginal desert-like lands, but also the more productive agricultural regions. Severe salinization is a very big problem in the western wheat-growing region. As a whole, Australia has about 5.7 million hectares (14 million acres) at risk for developing salinity problems, with more than 2.5 million hectares (6.2 million acres) already affected; this risk is predicted to rise to 17 million hectares (42 million acres) by 2050 (figure 8–24). Some estimates of lost agricultural production amount to more than $150 million (Australian) each year.

High levels of salt cause damages in the field, and also substantial damage in downstream ecosystems. In many parts of Australia, water has become brackish and saline and can no longer be used. Plants and animals, even native ones, cannot handle very high salt concentrations. Even if native vegetation in affected regions would be restored, some of these cleared areas could not recover. Urban areas are also affected, as salts can cause rust and corrosion of metals, lead to breaks in pipes, and destroy roads. There is no easy solution to salinization. Solutions will require management, technology, and strategy changes.

LOOKING AHEAD: GREENING THE JORDANIAN DESERT

Not long ago, the area near the Dead Sea was a salty, dusty mess. Irrigation mismanagement, drought, and overgrazing had left little behind but a few scraggly plants and salt crusts. The country of Jordan is currently more than 90% desert, has high temperatures, very little rain, and very little chance of creating productive farmlands. The people in the region are

Figure 8–23. In this saline seep salts accumulate on the surface (white areas), making soil unsuitable for growing and maintaining crops.

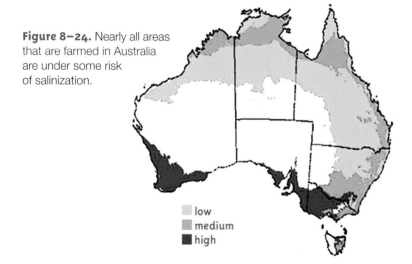

Figure 8–24. Nearly all areas that are farmed in Australia are under some risk of salinization.

Figure 8–25. The Re-Greening the Jordanian Desert initiative employs a permaculture system that helps remediate the salinity problems of neighboring villages by using mulches and shading.

very poor. However, a project in permaculture, started by an Australian, has worked to turn more than several thousand square meters into a green, lush growing area for fruit trees and crops (figure 8–25).

Permaculture combines agroforestry, sustainable farming and development, organic farming, agroecology, and social development. The overall idea is to help people be more productive and self-reliant.

The permaculture project in Jordan teaches participants that the first step in any self-sufficient system is to suppress evaporation while harvesting incoming water. This reduces the amount of salty groundwater rising in the soil profile, and reduces surface salinity.

The suppression of water evaporation can be done in several different ways. A farmer can place mulch over the soil

Figure 8–26. The green roof on Chicago's City Hall helps filter the air, reduce the urban heat island effect, and decrease the runoff from urban spaces.

surface, which both increases the moisture in the soil and decreases exposure to the sun's evaporating rays, lowering the accumulation of salts. The construction of sunken beds, which involves digging a pit so that water can flow down to where plants are growing, also helps to harvest incoming water. Rainwater has many fewer dissolved salts than groundwater, so plants have an easier time using rainwater. This technique may even capture enough water to leach out dangerous salts from the soil. The project also provides shaded tarps to place over crops to reduce evapotranspiration. The less evapotranspiration, the less water a plant needs to pull from the soil, reducing the chances of pulling up and accumulating salts in the topsoil.

This project has succeeded in combining various sustainable agricultural ideas, as well as reclaiming the salty desert land and bringing a better livelihood to the people of the region. It is a good example of how humans can adapt to the environment and find new solutions to difficult problems.

HAVE WE LEARNED ENOUGH FROM PAST MISTAKES?

All across the world people in cities and other developed areas, once so disconnected from farming and their soil, are discovering how important healthy soil is to their lives. In large cities in the United States and Europe, the number of rooftop and container gardens is increasing. There are also *green roofs* that use soil to combat air pollution and the urban heat island effect, including Chicago's City Hall (figure 8–26). This rooftop garden, installed in 2001, has

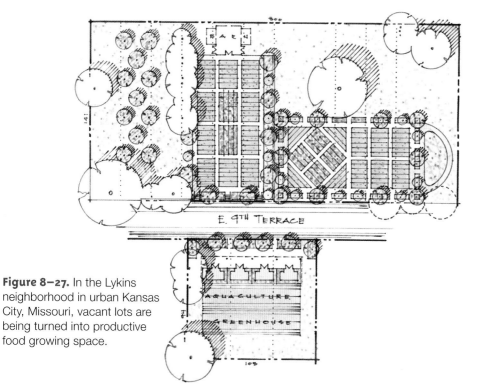

Figure 8–27. In the Lykins neighborhood in urban Kansas City, Missouri, vacant lots are being turned into productive food growing space.

more than 150 species of plants, producing oxygen for the city, reducing energy costs for air conditioning, and absorbing and re-using rainwater. In fact, soil scientists are highly active in designing new soil mixtures specifically for green roofs to provide the right amount of water retention and permeability to support dense vegetation without weighing too much. A healthy soil is the foundation of all the benefits a green roof provides—reduced runoff and erosion, filtration of pollutants, carbon dioxide storage, and energy efficiency.

Another trend is the creation of urban community garden programs. The Urban Farming Guys project, for example, is a sustainable farm in the Lykins neighborhood in urban Kansas City, Missouri (figure 8–27). Neighbors learn to cultivate healthy soil to grow food and flowers. They are using low-tech, high-yielding solutions in small spaces with a small budget. The neighborhood has also seen a decline in violent crime since the garden moved in. Several other cities and nonprofits organizations are reclaiming *brownfields* (urban sites that used to be industrial) for the purpose of starting community gardens and urban farms, providing jobs as well as food for local citizens.

Finally, more and more people are demanding organic food and local farmers' markets. Although advancing technology and chemical production led to increased crop yields, the advantages of intensive monocrop farming have been countered by many environmental costs, as detailed here and in chapter 6. Sustainable, organic farming focuses on soil and environmental health along with some social considerations. The U.S. Department of Agriculture defines organic farming as "an ecological production management system that promotes and enhances biodiversity, biological cycles, and soil biological activity" with strict regulation of the term *organic*. Most of the principles of organic farming emphasize biological soil additives and crop practices that lead to long-term soil health.

The practice of organic farming continues to become more and more popular with many farmers, gardeners, and consumers. A 2009 Worldwatch Institute report declared that land area under organic farming had increased by 150% since the year 2000, growing to 37.2 million hectares (80.8 million acres) worldwide. One of the strongest proponents of organic farming, Robert Rodale (1930–1990), claimed, "Feed the soil and it will feed the crop." Many proponents of organic farming now add "and it will feed the world."

Soils and Music

For centuries, soil has not only been important in literature and art, but also memorialized in songs. Soils can be an image or a metaphor or symbol for death, growth, or rebirth, and a powerful way to pass a message to another generation.

Woodie Guthrie's *Dust Bowl Ballads* (1940, RCA Victor) features songs about the hardship of the Dust Bowl, including "Dust Can't Kill Me."

Alison Krauss & Union Station's "Dust Bowl Children" (Rounder Records) is about the environmental mismanagement that led to the Dust Bowl, and how some families lost everything.

There are many other examples of song lyrics, including many country songs, folk songs, and even death metal songs, that invoke the imagery of soil. Can you find some?

The critical issue of sustainable, healthy food and crop production—organic or traditional—relies on understanding the soil as a whole. The crops must get enough nutrients and water. The soil biology needs to be healthy and diverse. The soil must be managed to avoid degradation from erosion, salinization, and compaction. BY TAKING CARE OF THE SOIL, SOCIETIES WILL BE ABLE TO TAKE CARE OF THEMSELVES.

SUMMARY

Today, we possess powerful new technologies and the ability to grow crops in many ways and then move food from productive areas into areas of need. We have modern medicine, instant global communication, and new knowledge about the workings of the world. Because of these things, each society's fate is more interconnected than ever before. The hope is that despite the challenges our world now faces, our individual societies will not fail as those of the Mesopotamians and others did, and that instead we will achieve global sustainability.

Educated citizens are making these changes as they realize that it is society's responsibility to make sustainable choices, since every decision affects people's health as well as our environment. Many people around the world, in both rural and urban areas, are rediscovering a vital connection to the soil. Franklin Delano Roosevelt said it best when he stated that, "A nation that destroys its soil destroys itself."

CREDITS

8–1, Lascaux Cave, French Ministry of Culture; 8–2, Tagulmoust; 8–3, Hieronymus Bosch, currently at Museo del Prado, Madrid, Spain; 8–4, Frank Watts; 8–5, Gary Simpson, *Common Ground 191*; 8–6, *from earth: stiftsland*, 1997, photo by Bruno Schneyer, Zeil am Main; copyright: herman de vries, Eschenau; 8–7, Prologue of the Code of Hammurabi, the Louvre, Paris; 8–9, National Archives; 8–11, Trazzler.com; 8–12, Google Earth, photo by Artemio Urbina; 8–13 and 8–14, NASA; 8–15, USDA-NRCS; ; 8–16, USDA; 8–17, Google Earth; 8–18, Báthory Péter; 8–19, gela21; 8–20, Trees for the Future; 8–21, Lambert; 8–22, U.S. Army; 8–23, MT Salinity Control Association and MT Bureau of Mines an Geology–MT Tech; 8–24, Australian Government; 8–25, Jordan Valley Permaculture Project, "Greening the Desert – the Sequel;" 8–26, copyright City of Chicago; 8–27, TheUrbanFarmingGuys.com. Chapter opener image, iStock.

Copyright © 2012. Soil Science Society of America, 5585 Guilford Rd., Madison, WI 53711-5801, USA. *Know Soil, Know Life*. David Lindbo, Deb A. Kozlowski, and Clay Robinson, Editors
doi:10.2136/2012.knowsoil.c8

GLOSSARY

ANTIBIOTIC Organic compound that in low concentrations is inhibitory to other organisms.

BIODIVERSITY The degree of variation of life forms within a given species, ecosystem, biome, or an entire planet. Often used as a measure of health of an ecosystem.

BLACK BLIZZARD A colloquial term for a dust-storm in the dust bowl of the south-central United States.

BROWNFIELD Abandoned or underused industrial and commercial facilities or sites available for re-use or redevelopment.

CAIRNS Human-made pile or stack of stones set up as a landmark or memorial.

CASH CROP A crop, such as tobacco, grown for direct sale rather than for livestock feed.

DEFORESTATION The permanent removal of trees until less than 10% of the forested land remains.

DESERTIFICATION Land degradation where the land area is becoming increasing dry and inhospitable to plants and animals.

EROSION (i) the wearing away of the land surface by rain or irrigation water, wind, ice, or other natural or anthropogenic agents that abrade, detach and remove geologic parent material or soil from one point on the earth's surface and de-posit it elsewhere; (ii) the detachment and movement of soil or rock by water, wind, ice, or gravity.

GEOPHAGY Deliberately ingesting clay or soil.

GREEN ROOF A roof of a building that is partially or completely covered with vegetation and a growing medium (soil), planted over a waterproof membrane.

PATHOGEN Any disease-producing agent, especially a virus, bacterium, or other microorganism.

PERMACULTURE An agricultural system or method that seeks to integrate human activity with natural surroundings so as to create highly efficient self-sustaining ecosystems.

SALINIZATION The process whereby soluble salts accumulate in the soil.

THERMALISM The therapeutic use of hot-water springs.

CHAPTER 9

CAREERS IN SOIL SCIENCE: DIG IN, MAKE A DIFFERENCE

Soil scientists work in a variety of areas and locations…it's usually not your typical job behind a computer all day. What are they doing? Where do their careers take them?

A forensic soil scientist finds himself in a quarry for a second time. He has sampled the soil at this site and from the trunk of a car where the murder weapon was found and is confident this is the same soil. The body must be somewhere.

A soil scientist is looking at the pattern of lead occurrences and other pollutants on a map of a neighborhood. She is trying to discern a pattern to make a hypothesis about the origins of the lead contamination. Did it come mostly from cars? Or, maybe house paint?

Another soil scientist is out in a soybean field in the Midwest. His job is to protect the health of both the growing crop and the groundwater by recommending fertilizer application rates and timing. He worries about how the recent drought is affecting our cropland and thinks he will do some research later on best practices under drought conditions.

A well-known soil scientist is speaking with a very large nongovernmental organization trying to bring health and nutrition to Africa. He argues that the key to ending famine is soil, but the message is strange to many. His colleague is already on the ground in Sierra Leone. She is helping residents of a village implement mixed cropping agroforestry practices that can help improve soil productivity.

There are thousands of soil scientists, and there are thousands of "typical days."

You've read, studied, and learned about the exciting world of soils and soil science. Is soil science a career for you? Ask yourself these questions: Are you fascinated with science? Enjoy working outdoors? Interested in how the ecosystem supports life on Earth? Do you want to make a difference in the environment—from soil conservation

CHAPTER AUTHORS

SUSAN CHAPMAN
DAVID LINDBO

to sustainable use of our land, water quality to wetlands, forests to deserts, and more? Are you energized talking about the importance of soil and its crucial role in sustaining life? Are you looking for days that are far from "typical"? THEN SOIL SCIENCE MAY BE THE CAREER FOR YOU!

This book has introduced the important aspects of soil, and the importance of soil to people and ecosystems. The concepts discussed here are fundamental for scientists who work with soils. Soil scientists deal with soils as a natural resource on the earth's surface. Soil scientists may work on soil formation, classification, and mapping. They may study the physical, chemical, and biological properties of soils. Or they may research all of these properties in relation to the fertility and management of soils.

Soil science, as we've seen, encompasses biology, ecology, and a variety of earth and other natural resource sciences. It intersects with geology and geography. It focuses on understanding, managing, and improving land and water. Soil science includes using chemistry, physics, microbiology, and mathematics. It also depends on high technology tools for soil exploration, analysis, data interpretation, and modeling of soil and landscape processes. Above all it integrates concerns for people, agriculture, and the environment.

SO THEN, WHAT DOES A SOIL SCIENTIST DO?

A soil scientist brings science and technology to bear on issues involving soil and water resources. Soil scientists are involved in soil and water management for a variety of purposes—from urban to agricultural to environmental—and play key roles in public and private decisions related to land use, soil, and water resources. They engage in research, teaching, consulting, lab work, field work, and many more activities. In the course of a career, a soil scientist might do any of the following:

- TEACHING—students in high schools, community colleges, universities, as well as farmers and community leaders;
- RESEARCH—from the field to the laboratory, using complex technology, GIS systems, statistics, and inter-related sciences;
- AGRONOMIC CONSULTING—working with farmers on how to best maximize crop production while ensuring the health and productivity of the soil;
- ENVIRONMENTAL CONSULTING—working with clients on issues such as water quality, site assessments, erosion, contamination, remediation, land use, wetland restoration, nutrient management, waste management, and much more;
- MAPPING—classifying soil types, based on soil properties, and mapping them through geo-encoding for appropriate land use;
- CONSERVATION—ensuring appropriate management strategies are used to conserve soil resources

WHO HIRES A SOIL SCIENTIST?

As a soil science graduate, you might find employment (depending on your level of education and experience) at universities and other educational institutions; with local, state, or federal government agencies; or in the private sector, with environmental or agricultural consultant firms, or with agribusiness companies (figure 9–1). Employment opportunities are available throughout the United States, North America, and across the globe.

So, for example, your job might be to:

- research strategies for containing radioactive waste at a nuclear storage facility,
- manage soils for crop production at a major agribusiness,
- teach a soil management course to agronomy students at your university,
- predict the effect of land management options on natural resources and prepare a report to be distributed by the federal agency where you work,
- advise land owners of capabilities and limitations of soils for various development projects as an independent consultant,
- recommend best-practice based soil management programs as an extension agent,
- conduct experiments to evaluate nutrient and water availability to crops as part of your research appointment at a university,
- advise clients on environmental issues such as water quality, site assessments, erosion, and wetland restoration as a consultant
- manage soils for mine reclamation and site restoration for the engineering firm where you work.

WHAT TYPE OF EDUCATION DO YOU NEED?

Soil scientists have earned at least a bachelor's degree from a major university. At many universities, two choices are available for specialized training in soils. One option prepares students to enter the agricultural sector as farm advisors, crop consultants, soil and water conservationists, or representatives of agricultural companies. The other option prepares soil scientists for careers in environmental positions dealing with water quality concerns, remediation of contaminants or for on-site evaluation of soil properties in construction, waste treatment and disposal, or recreational facilities.

Students of soil science in either track learn to identify, interpret, and manage soils in an environmentally responsible way. Laboratory and field coursework, which may involve soil judging and undergraduate research, is generally a component in core soil science courses. As a student majoring in soil science, you will take core courses in:

Skills, Knowledge, and Abilities a Soil Scientist Needs

Biology	Physics	Innovation	Reading comprehension
Chemistry	Work styles	Science (rules and methods)	Complex problem solving
Mathematics	Attention to detail	Active listening	Active learning
Education/training	Integrity	Critical thinking	Systems analysis
Computers and electronics	Initiative	Communication Skills	Writing
Communications and media	Dependability	Judgment and decision-making	Reasoning
Geography	Independence		Observation
	Persistence		
	Adaptability		

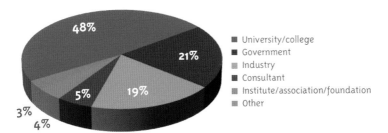

Figure 9–1. Where soil scientists work. More than 6,000 soil scientists are members of the Soil Science Society of America and work in all aspects of soil science, in many sectors. About 80% of these soil scientists live in the United States and Canada. There are also about 1,000 more soil scientists not reflected in this chart who are certified by SSSA and work in the consulting field.

- soils and land use management
- soil genesis, morphology, and classification
- soil biology and soil ecology
- soil chemistry and mineralogy
- soil fertility and nutrient management
- soil physics

In addition, you will also take some combination of supporting courses, such as:

- agricultural science
- biological and ecological sciences
- chemistry, mathematics, physics, and statistics
- communications
- geoscience and atmospheric science
- human health and land use
- technology and engineering
- water sciences

DO I NEED A CERTIFICATION OR A LICENSE TO BE A SOIL SCIENTIST?

Numerous professions involve certification, which is a voluntary credential, or licensing, which is mandatory. A license is generally required by a state government. Both certification and licensing have education and experience requirements and also require ongoing continuing education after receiving the certification/license.

For the soil science profession, the Soil Science Society of America (soils.org) offers a voluntary Certified Professional Soil Science certification. Individuals apply for the certification, which requires a certain number of years of experience in the profession, a minimum level of education (generally a bachelor's degree), passing required examinations, and providing letters of reference. A certification board reviews and approves or denies certification applications. If you are in private industry or consulting, this certification lets your clients know that you have met a professional standard.

As of 2012, the states of Maine, Virginia, New Hampshire, North Carolina, South Carolina, Minneapolis, Wisconsin, Tennessee, Texas, and North Dakota license soil scientists. Since other states may be initiating soils licensing, it is best to check with your state department of regulations and licensing to see what it requires.

The demand for soil scientists is expected to increase. According to one occupational report, the projection for 2010 to 2020 is 10% to 19% employment growth for soil and plant scientists. As with many professions, the higher the degree, the more opportunities are available, generally with greater options, responsibilities, and salaries.

Soil science is an important career, with substantial psychological benefits from knowing that you make a difference. A soil scientist contributes directly to the health of our environment through proper management of this essential resource. Soil scientists enjoy a variety of options for employment, as well as opportunities for advancement through additional education. The next section profiles working soil scientists. Read their stories as examples of where soil science can take you. This versatile career can't be represented with just a few biographies though—every soil scientist writes his or her own life story.

Famous Soil Scientists

Did you know …

- A recent Governor of Montana was a soil scientist?
- Charles Darwin is credited with the first scholarly treatment of soil-forming processes?
- Several soil scientists have won the World Food Prize?
- A soil scientist owns the company that built a thermal and electrical conductivity probe for NASA's Phoenix Scout Lander mission to Mars?
- Soil scientists can help solve crimes through soil forensics?
- Soil scientists use their knowledge of soil properties to help winemakers produce great wines?
- A former President of Bangladesh was a soil scientist?
- At least two Nobel prize winners have been involved in research related to soils?

CREDITS

Portions of the soil scientist career activities, skills inventory, and education requirements courtesy of USDA-NRCS and O*Net online. Occupational report statistics from O*Net Online Occupation Summary Report for Soil and Plant Scientists. Bill Shuster photos courtesy of Patrick Drohan and USEPA. Chapter opener image, M. Holzer.

Copyright © 2012. Soil Science Society of America, 5585 Guilford Rd., Madison, WI 53711-5801, USA. *Know Soil, Know Life.* David Lindbo, Deb A. Kozlowski, and Clay Robinson, Editors
doi:10.2136/2012.knowsoil.c9

A SERIES OF ADVENTURES!

LISA "DIRTGIRL" BRYANT

ASSISTANT FIELD MANAGER
BUREAU OF LAND MANAGEMENT | MOAB, UTAH

I have worked as a professional soil scientist for more than 20 years. Although I have managed soil, water, air, weed, and pesticide programs and I am now in a supervisory role in management, in my heart I will always be a soil scientist. I received my degrees at:

WASHINGTON STATE UNIVERSITY
B.S., General Agriculture with a minor in Statistics

UNIVERSITY OF CALIFORNIA–DAVIS
M.S., Soil Science

HOW LONG HAVE YOU BEEN A SOIL SCIENTIST?

I've been digging around in the yard and interested in bugs and flowers since I was a little kid, but I officially began my career as a soil scientist in 1989, working for the Tahoe National Forest, in Nevada City, California. I worked as a professional soil scientist for more than 20 years, but just last fall made the "leap" into management and supervision.

WHAT WAS YOUR CAREER PATH TO YOUR CURRENT POSITION?

I was hired for my first job on the Tahoe National Forest under their student coop program and worked part time while I finished my master's degree. I loved working in the foothills and forests of the northern Sierras, and stayed on the Tahoe National Forest for several years. In order to continue to advance and grow in my career, I took a job with the U.S. Bureau of Reclamation. In that job I worked in classifying lands for irrigation and working on water conservation projects throughout the Central Valley of California and southern coastal region. I returned to the Forest Service several years later to work on the Inyo National Forest in the Eastern Sierras and also learned about range, watershed management, and air quality. That led me to a job in Fort Collins, Colorado for the Arapaho Roosevelt National Forest and Pawnee National Grassland. Then I met a cute boy at a music festival in Colorado and applied for a job in Salt Lake City for the Bureau of Land Management (BLM) in order to get married. I was the statewide lead for Soil, Water, Air, Weeds, Pesticides, and the Grasshopper/Cricket Programs. Just last fall I moved to Moab as the Assistant Field Manager for the local BLM office. My career path has been a series of adventures based on living in interesting places, meeting new people, and following opportunities for advancement, personal, and professional growth.

WHAT PROJECTS ARE YOU WORKING ON NOW? WHAT INTERESTING PROJECTS HAVE YOU WORKED ON IN THE PAST?

In my current position as the Assistant Field Manager, Resources Division, I supervise a team of 13 people that work in wildlife, range, minerals, cultural history and archeology, oil and gas, and lands/realty programs. In just the last month we've worked on a uranium mining project, permitted filming for the new Lone Ranger movie, coordinated with partners on large-scale riparian restoration projects, and negotiated mitigation measures to help protect bighorn sheep, while still allowing for oil and gas exploration and providing

Field work in Utah describing and sampling soil.

"It's my backyard too, and I hope that the decisions that I make today will result in healthier landscapes for our children and grandchildren, but still provide for resources that we need."

for energy development. Each day poses new challenges and problems, and I work with a dedicated team of individuals that often have to work together and with other agencies or partners to help find solutions.

WHAT DO YOU FIND MOST INTERESTING ABOUT YOUR WORK?

There is always something new to learn! In addition to running into new challenges every day related to balancing land uses, like grazing and energy development, with protecting wildlife and riparian resources, I'm now learning more about people— what makes them tick, how they are motivated, how I can help them achieve their goals and work together better as a team. Not just the employees that work for me either—I work with a diverse community with many different views on how I should do my job. It's a whole new aspect of land management and a vital piece of the equation that I'm finding a little frustrating at times, but fascinating as well.

WHAT DO YOU LIKE BEST ABOUT YOUR CAREER?

Being able to make a difference: we ask a lot of our public lands—places to play and recreate, refuges for sensitive plants and animals, interpretation of our past and cultural heritage, providing energy, clean air and water. It's a difficult balance. It's my backyard too, and I hope that the decisions that I make today will result in healthier landscapes for our children and grandchildren, but still provide for resources that we need.

WHAT'S THE ONE THING YOU WOULD CHANGE ABOUT YOUR JOB/PROFESSION THAT WOULD MAKE IT BETTER?

I often tell people that "I love my job, I just wish there wasn't so much of it!" I just don't have enough time in the day to accomplish everything that I want to do. It would be nice to have the luxury of really spending the extra time to make sure that our work is done as the "absolute best we can do," not just the "best we can, given limited time/resources."

WHAT WOULD YOU SAY TO SOMEONE THINKING ABOUT BECOMING A SOIL SCIENTIST/LAND MANAGER?

Go for it! Science is always evolving— we learn more each day, the challenges and problems get more complex, so there's a lot of opportunity. While I think soils are fascinating and we don't know enough about them, hydrology, botany, cultural history, and renewable and sustainable energy are all interesting fields as well. Pick something that seems to resonate for you, and learn more about it and how it relates to the other resources. And, don't give up. Budgets are tight, but we need new talent and ideas and will hire good people when we get the chance.

WHAT ADVICE WOULD YOU GIVE FOR SUCCEEDING IN THIS FIELD?

Be mobile. Each new place I've been has been vastly different in terms of geology, soils, landforms, vegetation, and issues that needed to be addressed. I've gained a lot of my experience by being willing to move around and learn about different regions. What works in one may not in another, or it may with a few modifications, so moving around can help improve analytical and problem solving skills.

WHAT DO YOU LIKE DOING IN YOUR FREE TIME?

I hike, bike (mountain and road), paint with watercolors, attend music festivals and volunteer as a radio DJ at a local station, and I love to swim. Tonight I just enjoyed sitting on my front porch in the dark listening to the thunder and watching for lightning, in the first rainstorm that we've had in nearly three months.

A LIFELONG SCIENTIST!

BILL SHUSTER

RESEARCH HYDROLOGIST
U.S. ENVIRONMENTAL PROTECTION AGENCY

I am an environmental scientist in federal service, and I study urban soils and their use to manage stormwater runoff and provide other ecosystem services. I received my degrees at:

UNIVERSITY OF MICHIGAN—ANN ARBOR
B.S., Physics

THE OHIO STATE UNIVERSITY—COLUMBUS
Ph.D., Environmental Science

HOW LONG HAVE YOU BEEN A RESEARCH SCIENTIST?

I have been an environmental scientist all of my life! I took my first water sample at seven years of age, when I was trying out for swim lessons in 1972. They wondered why I had a test tube in my hand, and you can't imagine the interest this generated. Keep in mind that this was also the year that USEPA became a major force in improving environmental quality for our nation.

WHAT WAS YOUR CAREER PATH TO YOUR CURRENT POSITION?

After earning my degree and driving a lumber truck for about one year, I interned on a family farm in Maine, and I took that summer to figure out what I was going to do with a degree in physics. I wrote to USDA to see if there was any work that would bring soils and physics together. They wrote back and said "how about soil physics?" I had the good fortune of working as a technician in soil physics and soil ecology for seven years and learned some good skills that led me to an interdisciplinary Ph.D. program that integrated soil science, ecology, statistics, and civil engineering. My flexibility in working in the areas of soils and sustainable water resources management got me into USEPA as a research hydrologist.

WHAT PROJECTS ARE YOU WORKING ON NOW? WHAT INTERESTING PROJECTS HAVE YOU WORKED ON IN THE PAST?

I currently work in urban communities and take deep soil cores to understand how soils form and change in urban areas. I also make measurements to determine how fast water moves through these soils at different depths. Many cities are looking for ways to do something useful with

> Being able to communicate what you do is critical to making sure folks understand the importance of soils in their life. Bill also uses this part of his job to inspire students to ask questions and perhaps pursue career in soil science.

stormwater, and I say, let's see about letting it soak into the soil, and complete the urban water cycle. Yet, we first have to know about the qualities of these soils and what might make them useful to better manage our urban environments.

In the past, I have installed rain gardens (soils + plants) and rain barrels in a suburban neighborhood to see if people liked them, and if these practices soaked up or stored enough rainwater to improve the environment. I worked with an economist, chemist, ecologist, and lawyer. It is important to have a team to work on the tough problems in environmental protection.

WHAT DO YOU FIND MOST INTERESTING ABOUT YOUR WORK?

I like learning about the different soils around our nation, and understanding that every soil presents opportunities to help us manage the environment.

WHAT DO YOU LIKE BEST ABOUT YOUR CAREER?

I feel strongly that we all have a role in environmental management and sustaining our life support systems on Earth. So, every day I feel that I contribute to being a good steward of the earth and communicate this to my fellow citizens at every opportunity.

Using a compact, constant head permeameter to measure how fast water moves in the soil. He will use this information to help manage urban stormwater.

WHAT'S THE ONE THING YOU WOULD CHANGE ABOUT YOUR JOB/PROFESSION THAT WOULD MAKE IT BETTER?

I would rather be in the field 100% of the time, but that would not leave time for other important aspects of what a scientist does, which includes analyzing data, writing papers, and communicating with citizens and scientists.

WHAT WOULD YOU SAY TO SOMEONE THINKING ABOUT BECOMING A RESEARCH SCIENTIST? WHAT ADVICE WOULD YOU GIVE FOR SUCCEEDING IN THIS FIELD?

I would say "we need you!" Working as a technician can be tough work, but it is the best place to pick up field and lab skills and be paid for this work, and it will help you decide what it is that you would really like to do. There are many rewarding paths to take; do not prematurely jump into graduate school until you have a firm grasp of the tools of the trade.

WHAT DO YOU LIKE DOING IN YOUR FREE TIME?

I enjoy spending time with my family, playing electric and acoustic bass, dabbling in electronics, and practicing my photography skills.

MY DIVERSE LIFE AS A SOIL SCIENTIST

DAWN FERRIS

SOIL SCIENCE PROGRAMS COORDINATOR
SOIL SCIENCE SOCIETY OF AMERICA

I have been working as a soil scientist/hydrologist/environmental scientist since graduate school or for 22 years. I am a licensed Soil Scientist in Minnesota, and I am a Certified Professional Soil Scientist. I received my degrees at:

UNIVERSITY OF WISCONSIN—MADISON
B.S., Soil Science

UNIVERSITY OF MINNESOTA—TWIN CITIES
M.S., Soil Physics

UNIVERSITY OF MINNESOTA—TWIN CITIES
Ph.D., Forest Hydrology

WHAT WAS YOUR CAREER PATH TO YOUR CURRENT POSITION?

Graduate school, environmental consulting (where I worked for several firms and then owned my own consulting business), county government, academia, and now SSSA.

WHAT PROJECTS ARE YOU WORKING ON NOW? WHAT INTERESTING PROJECTS HAVE YOU WORKED ON IN THE PAST?

Currently I work on certification and licensing, including legislation. I also work with the Soils Certifying Board and the Council of Soil Science Examiners, focusing on education, examinations, and ethics. I also teach on-line continuing education courses for our certified professionals and others looking for continuing education credit (CEUs). Additionally I travel and speak to students, faculty, soil science organizations, etc. with the goal of facilitating communication throughout the soil science community.

As an environmental consultant I worked all over the United States on various large projects involving environmental review documents, wetlands, restoration, endangered species, and expert witness cases, to name a few. As a faculty member at The Ohio State University my most interesting trips and research were located in Iceland. In Scott County, Minnesota I was the county's first Natural Resources Program Manager and finished their first County Water Plan; I was an administrator for two watersheds and ran their Parks program. In the Parks program I facilitated the purchase and protection of numerous acres of future park land and green space.

WHAT DO YOU FIND MOST INTERESTING ABOUT YOUR WORK?

I have most enjoyed the travel that has been part of my career. The most notable place is Iceland, where I worked with the Icelandic Soil Conservation Service and learned about their challenges to reclaim/restore their degraded landscape. Iceland is a beautiful country with wonderful people that care deeply about their land and are focused on soil conservation. I have also enjoyed my time traveling around the United States and learning about soils in different regions and working with many different people and agencies to complete projects, while at the same time protecting our natural resources.

WHAT DO YOU LIKE BEST ABOUT YOUR CAREER?

I like that I am always doing different things and applying my knowledge in ways I never thought I would. I have met many interesting people, learned a lot, and increased my knowledge in many different ways by working with people in many different professions. One of the best things I have realized is that soil science is so much more than a science unto itself. It is, instead, a science that has applications far and wide and can take a person in many different directions if you are open to the challenge.

WHAT'S THE ONE THING YOU WOULD CHANGE ABOUT YOUR JOB/PROFESSION THAT WOULD MAKE IT BETTER?

I have enjoyed my career so far and wouldn't change my experiences. I think the profession needs to communicate better with the public and younger students regarding the many different things that we do and how soils are so vitally important to life. I believe that the opportunities available and the need for soil scientists will grow over time because soil science is integrated with so many other sciences, policy, and current global issues. Those that truly understand soils can provide unique and educated perspective on many complex problems that the global population faces today.

"I have most enjoyed the travel that has been part of my career. The most notable place is Iceland, where I worked with the Icelandic Soil Conservation Service and learned about their challenges to reclaim/restore their degraded landscape.

WHAT ADVICE WOULD YOU GIVE FOR SUCCEEDING IN THIS FIELD?

Be open to the possibilities and opportunities that this profession offers you and be willing to speak up and educate people about how important soils are. Most people don't realize the assets that a soil scientist can bring to the team. We are, by training, a multidisciplinary profession, and those skills can be applied to a multitude of areas—from urban issues to agriculture, to natural resources, and everywhere in between.

WHAT DO YOU LIKE DOING IN YOUR FREE TIME?

My dogs are my hobby when I am not working. I compete with them in conformation and performance events (lure coursing, racing, and obedience). I am also a judge for performance events and travel all over the United States and Canada.

NEVER STOP LEARNING!

MISSY HOLZER

HIGH SCHOOL EARTH SYSTEM SCIENCE TEACHER
CHATHAM HIGH SCHOOL, CHATHAM, NJ

I have been a science teacher for 26 years and currently am teaching 100+ students about all aspects of Earth and Space Science—which is dynamic and rewarding. I received my degrees at:

COOK COLLEGE (SCHOOL OF ENVIRONMENTAL AND BIOLOGICAL SCIENCES)
RUTGERS UNIVERSITY
B.S. Environmental Planning

THE COLLEGE OF NEW JERSEY
Science Education & Teaching, M.A.T.

RUTGERS UNIVERSITY
Physical Geography, M.S.

RUTGERS UNIVERSITY
Learning & Teaching Science Education, Ph.D. candidate

WHAT WAS YOUR CAREER PATH TO YOUR CURRENT POSITION?

My college degree required I take a variety of science courses, and after working in various positions in the horticultural fields, I was looking for something more meaningful. After taking a course in foundations of education, I decided on a career in science education, for which my undergraduate courses prepared me.

WHAT PROJECTS ARE YOU WORKING ON NOW? WHAT INTERESTING PROJECTS HAVE YOU WORKED ON IN THE PAST?

Currently I am the president of National Earth Science Teachers Association, an organization with more than 1,100 earth science educators from around the country. I am also on the Soil Science Society of America's K-12 Committee since 2006, which is a perfect fit with my enjoyment of the topic of soils, and my desire to ensure that the soil sciences are part of all science curriculums. Working at the state and national level on science curriculums has provided me with avenues for dissemination.

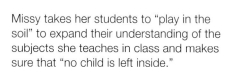

Missy takes her students to "play in the soil" to expand their understanding of the subjects she teaches in class and makes sure that "no child is left inside."

No matter where she goes—Hawaii, Bali, Africa—Missy looks at the landscape and soil for its beauty and relevance to her career. A termite mound can serve to demonstrate the importance of organisms in soil formation and show students that soils are alive and dynamic.

WHAT DO YOU FIND MOST INTERESTING ABOUT YOUR WORK?

I am passionate about my job and about science education, and the fact that my science field and my audience are dynamic makes my job extremely interesting. The fields within the earth sciences change on a continual basis as we learn more about the earth systems and how they interact. It is my role to develop ways to share this with my students at their levels. Each year my audience changes, and from one year to the next, my students bring with them personalities, abilities, and interests, all of which need to be considered when I design my lessons. I enjoy this aspect of my job immensely, especially when I can see the light bulb go off in the students' minds!

WHAT DO YOU LIKE BEST ABOUT YOUR CAREER?

Well-designed professional development makes all the difference in what is done in a classroom. I seek out engaging professional development during the summers and the school year, and the ones I enjoyed the most were my field experiences, which included going out to sea a few times on research vessels and collecting data in numerous places around the globe. These field experiences make their way back into my classroom through targeted lessons around the experience that promote an understanding about how we come to understand how the earth works through scientific investigations.

WHAT'S THE ONE THING YOU WOULD CHANGE ABOUT YOUR JOB/PROFESSION THAT WOULD MAKE IT BETTER?

I think most teachers would say they would have no problem giving up grading papers! Also, there is a local to national tendency to politicize (public) education by those who lack experience in what takes place in schools. Education as an "issue" is a huge distracter and becomes a bit wearing at times.

WHAT WOULD YOU SAY TO SOMEONE THINKING ABOUT BECOMING A TEACHER?

Teaching is extremely rewarding, although it can be tough for the first few years until the necessary skills are developed. Identifying a strong mentor can make a huge difference in how the first years of teaching proceed.

WHAT ADVICE WOULD YOU GIVE FOR SUCCEEDING IN THIS FIELD?

Never stop learning, and never stop seeking to improve what you do!

WHAT DO YOU LIKE DOING IN YOUR FREE TIME?

Currently I am a Ph.D. candidate working on a dissertation, and when that isn't usurping my time, I love being outside running, hiking, cycling, kayaking, and anything else with an aerobic aspect to it.

ENJOY THE CHALLENGE!

MATTHEW "MATT" DUNCAN

AGRONOMY SERVICES MANAGER
CROP PRODUCTION SERVICES

As a consulting soil scientist/agronomist I work to prevent, solve, and correct issues associated with soils and plant growth in the industries of production agriculture, mining, oil and gas exploration, and transportation and private lands. I have been working in agronomy/soil science for 14 years am a Certified Professional Soil Scientist and a Certified Crop Adviser/Certified Professional Agronomist. I received my degrees at:

UNIVERSITY OF ILLINOIS—URBANA-CHAMPAIGN
B.S., Soil Science

PURDUE UNIVERSITY, LAFAYETTE, IN
M.S., Soil Science

WHY DID YOU CHOOSE TO BECOME CERTIFIED? WHAT VALUE DID IT ADD TO YOUR CAREER?

Certification provides employers, clients, and farmers a quick and easy way to know you have a particular level of knowledge, understanding, and education. It is also important when issues are brought into the legal system to establish yourself as an expert.

WHAT WAS YOUR CAREER PATH TO YOUR CURRENT POSITION?

While in college, during summer breaks I worked in a soil testing laboratory as a technician and as a crop scout identifying crop pests in the field. After college I became a research agronomist and started and ran a 300-acre agricultural research farm testing product effectiveness and conducting training and education tours for farmers and ag retail salesmen. I then moved into the seed industry as a technical agronomist. For the last half of my career I have been working as a technical soil scientist assisting agencies and industry in protecting agricultural lands during energy industry construction projects, such as interstate oil and gas pipeline construction.

Matt is outstanding in his field.

WHAT PROJECTS ARE YOU WORKING ON NOW? WHAT INTERESTING PROJECTS HAVE YOU WORKED ON/LED IN THE PAST?

One of the more unique projects I have worked on in the past was in New Jersey. The soils this construction project crossed included acid producing soils. These are soils that, when exposed to air and moisture produce sulfuric acid! The challenge was the soil had to be dug up to remove and replace a gas pipeline that passed through farmland, streams, and woods. Special testing, handling, and storage procedures had to be created to manage the soil. I currently work with a large agricultural retailer to educate and train salesmen and agronomists.

WHAT DO YOU FIND MOST INTERESTING ABOUT YOUR WORK?

I most enjoy the challenge of determining the cause of an issue and then creating viable solution options to correct the problem. I also enjoy working as an expert witness. Soil science is an incredibly complex science that includes aspects of practically all the sciences. If you really focus on your career, learn the science well, and always strive to continue learning, you will be an expert in a field that not many people really understand.

WHAT DO YOU LIKE BEST ABOUT YOUR CAREER?

I enjoy working with people from a wide range of backgrounds—farmers, engineers, geologists, government agencies, biologists, crop consultants, and so on—and the challenge of getting them to understand the importance of soil and soil management and its influence on plant growth, food production, and a healthy environment.

WHAT'S THE ONE THING YOU WOULD CHANGE ABOUT YOUR JOB/PROFESSION THAT WOULD MAKE IT BETTER?

I really don't know what I would change. Soil science is actually a very dynamic profession. I get to travel throughout the country for projects. I get to work both in the field and in the office, so neither becomes old. I get to meet and work with people from a wide range of backgrounds, and each issue or project is just different enough to offer up a new challenge in one form of another.

WHAT WOULD YOU SAY TO SOMEONE THINKING ABOUT BECOMING A SOIL SCIENTIST/AGRONOMIST? WHAT ADVICE WOULD YOU GIVE FOR SUCCEEDING IN THIS FIELD?

First, remember that all the sciences (chemistry, biology, physics), as well as some math and writing skills are important to any science-based field, especially soils. Next, pick a good college, one with not only soils classes, but with soil science as a degree major. When you look for summer jobs prior to graduating college, try to find something that relates to soils. And finally, be willing to work your way up; don't assume you are entitled to anything. A good education and attitude are the foundation, but you will still need to prove you deserve the job, promotion, or raise through your actions and successes.

WHAT DO YOU LIKE DOING IN YOUR FREE TIME?

I have always enjoyed fishing during my free time. I find it to be both relaxing and exciting—when they are biting.

Soil testing and early stage diagnostics of plant growth are an integral part of the training of professionals that Matt does in his current job. Working with the heavy machinery is a bonus.

SOLVING PROBLEMS, RECLAIMING LAND

BRUCE BUCHANAN

ENVIRONMENTAL CONSULTANT
BUCHANAN CONSULTANTS, LTD.

I consider myself both a Soil Scientist and an Environmental Scientist and have worked in the field of soil science since 1968! Most of my work has been in Forest Ecology or Mineland Reclamation. I currently design reclamation for surface coal mines in the arid Southwest. I am a Certified Professional Soil Scientist and a Certified Professional Soil Classifier. I received my degrees at:

UNIVERSITY OF UTAH
B.S., Botany

UNIVERSITY OF UTAH
M.S., Plant Ecology

MONTANA STATE UNIVERSITY
Ph.D., Forest Soils

WHAT WAS YOUR CAREER PATH TO YOUR CURRENT POSITION?

I've had a diverse career! I started my career as a summer employee for the Forest Service. After completing graduate work in 1971, I was a Professor at New Mexico State University for 20 years. Then, I started a consulting firm in 1991, Buchanan Consultants, where I am today.

WHAT PROJECTS ARE YOU WORKING ON NOW? WHAT INTERESTING PROJECTS HAVE YOU WORKED ON/LED IN THE PAST?

I design reclamation mainly. Reclamation is the process of revegetating disturbed lands. Disturbances result from activities like mining, highway construction, drilling, and fire. The objective is to protect the land from erosion and establish sustainable vegetation. Right now I'm working on salt movement in reclaimed soils for the oil and gas industry. Most of my work has been in the field of reclaiming mine lands. One project, at La Plata Mine, Northwest New Mexico, won numerous awards including a national award for the Best Reclamation in the United States. I have worked on projects in most of the western states. Reclamation has been conducted at high elevation (over 10,000 ft), forestry, rangeland and semi-arid deserts. I'm also involved in conducting a river survey of both the Animas and San Juan River and soil surveys for the White Mountain Apache Indian Reservation.

WHAT DO YOU FIND MOST INTERESTING ABOUT YOUR WORK SOIL SCIENTIST?

It is always interesting and rewarding to solve problems and make a difference in the field of soils so we have a cleaner environment. For example, some of the interesting projects I've been involved with are working on plans to close a copper mine in New Mexico. And, I've served as an expert witness, testifying on salt movement in the reclamation of oil and gas drilling locations and also as an expert witness with my soil science and reforestation expertise in a case involving logging practices in Vermejo Park, New Mexico.

"It is always interesting and rewarding to solve problems and make a difference in the field of soils so we have a cleaner environment."

WHAT DO YOU LIKE BEST ABOUT YOUR JOB AND/OR WHAT DO YOU LIKE BEST ABOUT BEING IN YOUR PROFESSION?

As a soil scientist, you are always learning!

WHAT'S THE ONE THING YOU WOULD CHANGE ABOUT YOUR JOB/PROFESSION THAT WOULD MAKE IT BETTER?

Making decisions on science and less on politics.

WHAT WOULD YOU SAY TO SOMEONE THINKING ABOUT BECOMING A SOIL SCIENTIST? WHAT ADVICE WOULD YOU GIVE FOR SUCCEEDING IN THIS FIELD?

I would advise students to be open-minded and not be set on ONE subject.

WHAT DO YOU LIKE DOING IN YOUR FREE TIME?

Fly fishing!

CHAPTER 10

SUMMARY AND PERSPECTIVES

CHAPTER AUTHORS

DAVID LINDBO
DEB KOZLOWSKI

Soils are everywhere and are dynamic, diverse, exciting, and beautiful. This book has introduced you to all their complexities and explained why soil should definitely *not* be called dirt! Soil is an overlooked, taken for granted, often abused, natural resource without which life as we know it would not exist.

We have covered the physical, biological, and chemical (chapters 2, 3, and 4, respectively) aspects of soils and demonstrated the environmental importance of soil to humans (chapter 6), as well as Earth's ecology (chapter 7). We have delved into the importance of soil to human history and culture (chapter 8) as well. You just can't appreciate how soils fit into our lives and the entire ecology of the planet without an understanding of the basic properties of soils. Perhaps all this may lead you to consider the vast array of careers available that involve soils (chapter 9)! The science of soil is complex, perhaps much more so than what you originally thought. It requires a strong understanding of the major fields of science (table 10–1) and is related to the liberal arts as well (chapter 8). We end here with some perspectives on the interrelationship of soils and our environment.

Let's look at the environment around us from the perspective of a working soil scientist, applying our knowledge of soils and soil processes to potential environmental effects on a landscape. Many soil scientists work on developing management plans in agricultural environments, like this landscape in the U.S. Midwest (figure 10–1). This is a beautiful rural landscape, and that is what most people driving across the countryside see. What does the soil scientist see? If the crop fields have a silty textured surface, the water *erosion* potential would be high since silty textured soils are more easily eroded than sandy or clayey textured soils (chapter 2). The riparian zone and filter strip (*best management practice*, BMP, chapter 6) at the base of the field would help slow runoff and remove some of the soil suspended in the runoff water. Using no-till practices or leaving more crop residue (chapter 6) could further reduce the amount of erosion. The increased organic matter (chapters 2, 3, 4, and 6) improves the soil structure by aggregating soil particles, thereby improving infiltration and lowering runoff. It also keeps the soil ecosystem healthy and diverse. In addition, the cover left by the crop residue helps dissipate the energy of raindrop impacts and reduces erosion potential.

As with all agricultural crops, some fertilizer is needed, but it should only be applied in the amount the crop needs based on a soil test (chapter 4). If a *crop rotation* such as legume–corn is in place, some of the corn's nitrogen (N) needs will be met with the residual N from the legume (chapters 3, 4, and 6). Excess N may leach as nitrate (NO_3^-) to groundwater and be carried toward the *riparian buffer*. If the

Figure 10–1. Looking at this landscape from the perspective of a soil scientist, the beautiful rolling hills suggest potential for erosion downslope. *Cover crops*, *contour planting*, and *conservation tillage* are best management practices (BMPs) that can help. Also, water in the landscape drains to the stream in the lower part of the photo, with the potential to carry excess nitrate (NO_3^-) and other nutrients to the stream. A soil scientist, through the development and implementation of a nutrient management plan, can help solve this potential problem. Filter strips and riparian buffers are BMPs that can further help to prevent excess movement of nutrients into the stream. The soil profiles shown remind us that the soil scientist investigates both the soil profiles and the landscape to understand the environment. The profiles at the base of the field in the riparian zone show signs of being saturated and anaerobic for part of the year. The NO_3^- may be denitrified if the NO_3^- containing shallow groundwater moves through the riparian zone under anaerobic conditions with adequate amounts of organic carbon present.

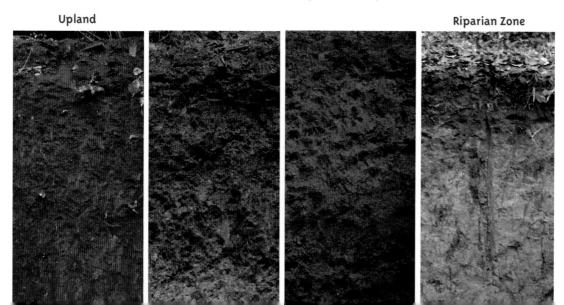

Table 10–1. Soil principles related to general fields of study.

Physics	Chemistry	Biology
water retention	fertility	organic matter
water movement	acidity	decomposition
oxygen diffusion	alkalinity	micro-organisms
heat transfer	ion exchange	macro-organisms
density	elemental form	root interactions
porosity	nutrient holding	nitrogen cycle
electrical conductivity	pesticide reactions	phosphorus cycle
particle size	trace elements	sulfur cycle
compactibility	source minerals	pathogens
shrink–swell	soil color patterns	soil color patterns

groundwater is anaerobic, as the nitrate (NO_3^-) enters the riparian zone some of it will be denitrified (chapters 3 and 4). The effectiveness of the riparian buffer for denitrification depends on oxygen content, which can be assessed by looking at the soil profile for redoximorphic features (chapter 2). Where the redoximorphic features are observed it is likely that the soil is likely saturated and anaerobic for part of the year. This means denitrification can potentially occur, especially since tree roots can provide the carbon for the microbial process (chapter 3). The riparian zone will also trap any phosphorus (P) and other chemicals adsorbed onto soil particles (chapters 4 and 6). Of course, the trees planted in the riparian zone should be native species, which are adapted to the biome (chapter 7).

A soil scientist considers all of these things when evaluating a situation with optimal land management in mind. If you have an opportunity to take a drive through the countryside or fly over agricultural fields in an airplane, take in the view with the mind of a soil scientist. There is a sophisticated, ever-evolving science that helps us manage our lands for sustainability and continued capacity to supply our society with food, fuel, feed, and fiber.

Consider a similar landscape but in an urban or suburban setting. Can our soil knowledge still be used for better environmental management? Beginning with the lawns so common in many areas, fertilizer use could cause problems with water quality. The best way to minimize the problems is to test the soil and apply fertilizer based on the results and recommendations (chapter 4) (figure 10–2a). Rain gardens can prevent storm water runoff from carrying lawn chemicals directly to water bodies (figure 10–2b).

Lawn fertilizers are not the only source of nutrients from a developed area. Human wastewater is often treated and dispersed into the soil effectively (chapter 6) (figure 10–2c). Aerobic soil treats most of the pathogens, but nutrients can still pose a problem. Much of the phosphorus (P) in wastewater is adsorbed by fine-textured soils but can leach in sandy soils or soils with low iron contents (chapters 2 and 4).

As in agricultural environments, soil scientists need to understand the soil properties in urban and suburban areas to understand the potential for negative environmental effects of excess nutrients (figure 10–2d). If the nitrogen (N) in the wastewater is in the ammonium (NH_4^+) form, it may be held in the soil on exchanges sites, but if the soil is aerobic, nitrogen (N) is likely in the nitrate (NO_3^-) form and is easily lost by leaching to groundwater. As the groundwater and nitrate (NO_3^-) move into the anaerobic riparian zone, the nitrate (NO_3^-) may denitrify and have limited overall environmental effect (chapter 3 and 4).

In developed areas storm water and sediment from construction sites are a major source of pollution as well. Erosion control strategies, such as silt fences can help (figure 10–2e). Riparian buffers are also used in suburban areas to help mitigate this problem in the same way they reduce the effects of erosion from an agricultural field: by trapping sediment and slowing the overall flow of runoff (figures 10–2f and 10–2g).

Unfortunately, soils, whether in developed or natural areas, may become contaminated with pollutants, such as oil and gasoline. There are several ways to deal with these and other contaminants. One method of *bioremediation* uses our knowledge of the carbon and nitrogen cycles along with naturally occurring soil microbes (chapter 3). This method involves adding fertilizer—nitrogen (N) and phosphorus (P) in particular—to the soil to increase the biological activity of the microbes. As their activity increases the microbes consume the carbon in contaminants and release it as carbon dioxide. This method was shown to be successful in mitigating the effects of the Exxon Valdez oil spill in Alaska. Similar effects can be seen by planting and fertilizing select species such as tall fescue and ryegrass, whose roots encourage the growth of microbes in the rhizosphere (*phytoremediation*). Knowledge of soil microbial processes (chapter 3) and soil fertility (chapter 4) are critical to the success of bio- and phytoremediation.

Soils and soil microbes are also important to human health (chapter 8). Investigations into soil microbes continue to turn up new beneficial pharmaceuticals.

We've seen examples of how soil management can address existing environmental issues. Knowledge of soil can also be used in land use planning to help prevent these issues from occurring. Recall that the soils on a landscape are formed through the interaction of the five factors of formation, or CLORPT (chapter 2). Using this knowledge we can predict what type and properties of soil we are likely to find on that landscape (figures 10–3 and 10–4) so that we can describe the soil (chapter 2)

a)

Figure 10–2. How do soil scientists view the landscape in a suburban or urban setting? Here too there are management challenges and solutions related to maintaining the health of the environment. (a) Soil testing is a good place to start, especially when lawn fertilizers are to be applied. (b) Rain gardens are a way to manage storm water runoff better. Runoff from impervious surfaces such as roofs and driveways collects in the depression of the rain garden and infiltrates into the soil, rather than flowing directly to storm drains that lead to nearby water bodies. (c) A septic system drain field makes a pattern of green stripes where there are more nutrients and water available to the lawn. Trees remove some nutrients from the wastewater, as will the grass over the drain field. The soil below the drain field will further treat the wastewater by physical filtration and biochemical processes that remove pathogens and some additional nutrients.

b)

c)

d)

Figure 10-2. Continued. (d) (from left) Well drained, to moderately well drained, to poorly drained soils. Nitrate (NO_3^-) moves through the well drained soil to groundwater, where it flows towards the moderately well to poorly drained soil. As the NO_3^- moves into zones where the soil is anaerobic, the NO_3^- is biochemically converted to dinitrogen gas (N_2) provided there is a carbon source (food) for the microorganisms. (e) This silt fence has been installed to prevent soil from being carried away with storm water. (f) A grassed filter strip along a (g) riparian zone traps sediment and slows runoff from nearby streets and lawns.

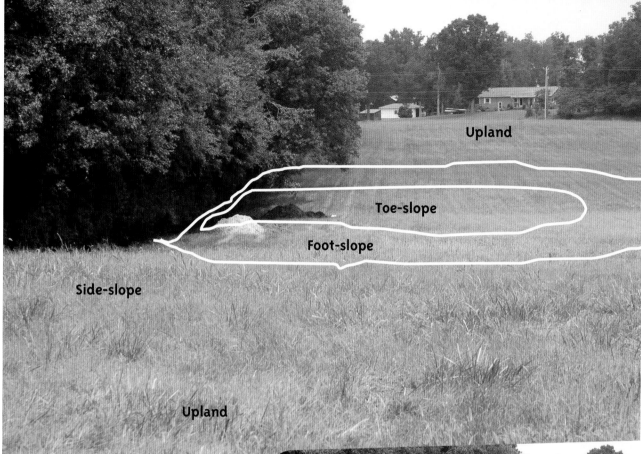

Figure 10–3. Typical field in the Piedmont of North Carolina. The low-lying area, or toe-slope, collects surface water and has a shallower water table than the adjacent uplands. Based on the bedrock geology and mineralogy we can predict that the soils in the low area will contain more base cations and are likely to have some shrink–swell clay.

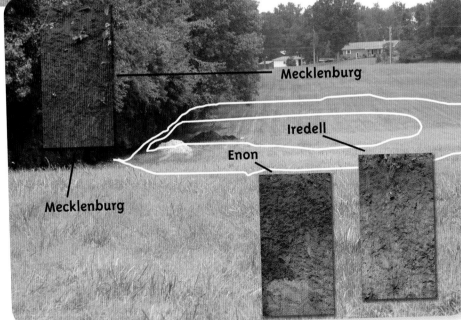

Figure 10–4. Soil map of the same field. As predicted, the soil in the low area is an Iredell Series, and the soil in the upland is a Mecklenburg Series. The Enon Series has properties intermediate between the Iredell and Mecklenburg Series. See table 10–2 for the complete soil descriptions of the Mecklenburg and Iredell Series.

Table 10–2. Profile descriptions for Iredell and Mecklenburg soils shown in figure 10–4.

Mecklenburg Series	Fine, mixed, active, thermic Ultic Hapludalfs
Ap	0–20 cm; reddish brown (5YR 4/4) loam; granular structure; friable; many fine roots; common fine pores
Bt1	20–42 cm; yellowish red (5YR 4/6) clay; subangular blocky structure; firm, moderately sticky, moderately plastic; common fine roots; few fine pores
Bt2	42–62 cm; yellowish red (5YR 4/8) clay; common fine distinct brownish yellow (10YR 6/6) mottles; subangular blocky structure; firm, moderately sticky, moderately plastic; few fine roots; few fine pores
BC	62–90 cm; yellowish red (5YR 4/8) clay loam; common medium faint reddish yellow (7.5YR 6/6) and common fine distinct yellowish brown (10YR 5/4) mottles; weak subangular blocky structure; firm, moderately sticky, moderately plastic; common fine lenses of gray clayey saprolite
C	90–150 cm; mottled yellowish red (5YR 4/8), reddish yellow (7.5YR 6/6) and yellowish brown (10YR 5/4) highly weathered saprolite that has a sandy clay loam texture; massive; friable

Iredell Series	Fine, mixed, active, thermic oxyaquic Vertic Hapludalfs
Ap1	0–12 cm; dark grayish brown (2.5Y 4/2) sandy loam; granular structure; friable, slightly sticky, slightly plastic; many fine and medium roots
Ap2	12–17 cm; dark grayish brown (10YR 4/2) loam; granular structure; friable, slightly sticky, slightly plastic; many fine and medium roots
Btss1	17–27 cm; brown (10YR 4/3) clay; angular blocky structure; very firm, very sticky, very plastic; many fine and medium roots; common slickensides
Btss2	27–50 cm; brown (10YR 4/3) clay; angular blocky structure; very firm, very sticky, very plastic; many fine and medium roots along faces of peds; common slickensides and pressure faces
Btg	50–60 cm; dark grayish brown (2.5Y 4/2) clay; angular blocky structure; common fine distinct yellowish brown (10YR 5/4) redox concentrations; very firm, very sticky, very plastic; many fine and medium roots, mostly along faces of peds; few fine pores
BC	60–67 cm; olive (5Y 4/3) loam; common very pale brown (10YR 7/3) and few distinct dark grayish brown (2.5Y 4/2) and black (N 2/0) mottles; angular blocky structure; firm, moderately sticky, moderately plastic; many fine and medium roots along faces of peds; few medium pores; common soft dark grayish brown and black saprolite
C1	67–80 cm; finely mottled dark greenish gray, very pale brown, and yellowish brown loam; 80 percent saprolite that crushes easily; many fine roots
C2	80–110 cm; finely mottled dark greenish gray, very pale brown, black and yellowish brown sandy loam; 90 percent saprolite that crushes easily; many fine roots
C3	110–155 cm; finely mottled dark greenish gray, yellowish brown, black and very pale brown sandy loam; 90 percent saprolite that crushes easily

and develop maps, using existing county soil surveys, or the Web Soil Survey (chapter 5). These maps can help us determine suitable land uses and those that would carry greater risks.

With that information we can identify possible BMPs to mitigate these risks. For example, on the open field and overlain soil map (figure 10–4, table 10–2) with the profile descriptions included (table 10–1), we can make several interpretations. The Iredell Series has a Btss horizon with redoximorphic features. The soil has a clay accumulation (Btss) in the subsoil that is also dominated by *shrink–swell clays*. It has a water table at 75 centimeters, indicated by the <2 chroma redoximorphic features (redox depletions). The high chroma redox features (redox concentrations) at 50 centimeters indicate that the water table may rise higher in the profile.

Given this description and classification the Iredell Series would have limitations for houses with basements and septic systems (wetness and foundation issues would arise from the shrink–swell clay). On the other hand, the Mecklenburg Series also has an accumulation of clay in the subsoil (Bt) but lacks the shrink–swell clay (no Btss). It has a deep water table, based on the lack of redoximorphic features. Therefore, the Mecklenburg Series would have few limitations. From this information, where would you choose to build a house?

Soil surveys can also be used to find locations for more esoteric purposes. Suppose you are a potter and want to use native clay in your art. Using a soil survey and looking for soils with a thick, high clay Bt horizon that is low in sand and coarse fragments would help you narrow your search. For many applications, once you know the important or limiting soil properties, your knowledge of soils and soil surveys can point you in the right direction for a particular use.

Above all, knowledge of soils and soil processes can help ensure that we produce enough food, fiber, fuel, and shelter to feed our growing world population. Soils are everywhere and are dynamic, diverse, exciting, and beautiful. We will have come full circle when you realize: **if you know soil, you know life, and if there were no soil, there would be no life!**

CREDITS

10–1, landscape, T. McCabe, USDA-NRCS, soils, D.Lindbo; 10–2a, urban soil sampling, USDA-NRCS and U.S. Forest Service; 10–2b, City of Lincoln, Nebraska Watershed Management Division; 10–2c, D. Lindbo; 10–2d, D. Lindbo; 10–2e, USEPA; 10–2f, 10–2g, 10–3, 10–4, D. Lindbo; chapter opener image, iStock. Table 10–1 adapted from H.J. Kleiss.

Copyright © 2012. Soil Science Society of America, 5585 Guilford Rd., Madison, WI 53711-5801, USA. *Know Soil, Know Life*. David Lindbo, Deb A. Kozlowski, and Clay Robinson, Editors
doi:10.2136/2012.knowsoil.c10

GLOSSARY

BEST MANAGEMENT PRACTICE (BMP) Any of a group of practices that conserve soil and water resources.

BIOREMEDIATION The use of biological agents, such as bacteria or plants, to remove or neutralize contaminants from polluted soil or water.

BUFFERS (BUFFER STRIP) Strips of perennial plants, such as grasses, along borders of fields or water bodies.

CONSERVATION TILLAGE Any tillage sequence, the object of which is to minimize or reduce loss of soil and water; operationally, a tillage or tillage and planting combination which leaves a 30% or greater cover of crop residue on the surface.

CONTOUR PLANTING Performing the tillage operations and planting on the contour within a given tolerance.

COVER CROPS Close-growing crop that provides soil protection, seeding protection, and soil improvement between periods of normal crop production, or between trees in orchards and vines in vineyards. When plowed under and incorporated into the soil, cover crops may be referred to as green manure crops.

CROP ROTATION A planned sequence of crops growing in a regularly recurring succession on the same area of land, as contrasted to continuous culture of one crop or growing a variable sequence of crops.

EROSION (i) The wearing away of the land surface by rain or irrigation water, wind, ice, or other natural or anthropogenic agents that abrade, detach, and remove geologic parent material or soil from one point on the earth's surface and deposit it elsewhere; (ii) the detachment and movement of soil or rock by water, wind, ice, or gravity.

PHYTOREMEDIATION A bioremediation process the uses plants to remove or neutralize contaminants from polluted soil or water.

REDUCED TILLAGE A tillage system in which the total number of tillage operations preparatory for seed planting is reduced from that normally used on that particular field or soil.

RESIDUE MANAGEMENT The operation and management of crop land to maintain stubble, stalks, and other crop residue on the surface to prevent wind and water erosion, to conserve water, and to decrease evaporation.

RIPARIAN BUFFER A vegetated area (type of buffer strip) near a stream, usually forested, which helps shade and partially protect a stream from the impact of adjacent land uses. Riparian buffers have become common conservation practices designed to improving water quality and reducing pollution.

SHRINK–SWELL CLAYS Clay minerals that expand greatly when wet or saturated, e.g., montmorillinite, bentonite.

SOIL SERIES The lowest category of U.S. system of soil taxonomy. A soil series is named for the area in which it was first mapped. A soil series is based on specific morphological, physical, and chemical properties that make it unique. It is equivalent to the species level in Linnaean classification.

If you KNOW SOIL, you KNOW LIFE, and if there were NO SOIL, there would be NO LIFE!